高校英语选修课系列教材

APPROACHING CARBON PEAKING AND CARBON NEUTRALITY

双碳通识英语教程

主　编　刘长江
副主编　贾　军
编　者　吴　婧
　　　　程义新
　　　　许　阳

清华大学出版社
北　京

内 容 简 介

本书紧密围绕国家"双碳"战略,以英语为载体,兼顾外语教学的工具性和人文性,全面介绍"双碳"领域的重要概念和相关知识,涵盖碳基生命、碳中和、碳达峰、碳足迹、碳市场、碳交易、新能源、绿色材料、绿色交通、绿色生活等方面的通识内容,旨在提升大学生对我国"双碳"战略、绿色发展、构建人类命运共同体的认知、认同和实践能力。本书配有教学课件和课后练习答案,读者可登录www.tsinghuaelt.com免费下载使用。

本书可作为高等院校本科生和研究生的能源英语类通识课程教材使用,也可作为"双碳"领域从业人员和广大英语学习爱好者拓展"双碳"通识知识、提高英语综合技能的参考书。

版权所有,侵权必究。举报: 010-62782989, beiqinquan@tup.tsinghua.edu.cn。

图书在版编目(CIP)数据

双碳通识英语教程 / 刘长江主编. —北京:清华大学出版社,2024.8
高校英语选修课系列教材
ISBN 978-7-302-66382-9

Ⅰ.①双… Ⅱ.①刘… Ⅲ.①节能—英语—高等学校—教材 Ⅳ.①TK01

中国国家版本馆 CIP 数据核字(2024)第 108757 号

责任编辑:刘　艳
封面设计:子　一
责任校对:王荣静
责任印制:刘海龙

出版发行:清华大学出版社
　　　　网　　　址:https://www.tup.com.cn,https://www.wqxuetang.com
　　　　地　　　址:北京清华大学学研大厦A座　　邮　　编:100084
　　　　社　总　机:010-83470000　　　　　　　　邮　　购:010-62786544
　　　　投稿与读者服务:010-62776969, c-service@tup.tsinghua.edu.cn
　　　　质量反馈:010-62772015, zhiliang@tup.tsinghua.edu.cn
印 装 者:涿州市般润文化传播有限公司
经　　销:全国新华书店
开　　本:185mm×260mm　　印　张:13　　字　数:217千字
版　　次:2024年8月第1版　　　　　　　　印　次:2024年8月第1次印刷
定　　价:58.00元

产品编号:106370-01

前　言

　　2021年9月，中共中央国务院在《关于完整准确全面贯彻新发展理念做好碳达峰碳中和工作的意见》中提出建设碳达峰、碳中和（以下简称"双碳"）人才培养体系，鼓励高等学校增设"双碳"相关学科专业和课程。2022年10月，党的二十大报告再次提出积极稳妥推进"双碳"重大战略决策，加快实现生产生活方式绿色变革，构建人与自然和谐共生的现代化。在此背景下，开展"双碳"领域通识教育与实践培养，将"双碳"理念与实践融入人才培养体系，支撑新时代产业发展人才需求目标，是高等学校必须担当的历史使命和社会责任。

　　为更好地服务新时期国家人才发展战略，响应新文科跨界跨学科整合的发展范式和新工科可持续发展理念与思维，满足大学生成长成才需要，南京航空航天大学双碳通识英语课程团队在深入学习"双碳"国家战略、充分领会《新文科研究与改革实践项目指南》的基础上，融合先进外语教学理论，编写了此教程，力求更好地培养具有国际视野、生态思维、文化自信的青年科技人才，支撑国家高质量人才培养。

编写理念与特色

一、体现新文科建设的核心理念——交互与融通

　　本教材介绍了碳基生命、碳中和、碳交易、新能源与新材料、绿色交通与绿色生活等领域的通识类知识，将"双碳"领域的新理念、新技术融入大学英语的课程中，提供契合现代社会需求的科学新知识和学术视角，培养学生生态节能的环保意识和文明共鉴的思辨能力，拓宽学生开放包容的国际视野，在人才培养、知识边界拓展和大学英语课程重构等方面实现突破。因此，本教材是建设新文科背景下大学英语课程改革的新尝试。

二、依据具有中国特色的外语教学理论——产出导向法

本教材中的模块设计理念依据的是产出导向法的"驱动—促成—评价"教学流程，突出产出导向、文道相融，为国家培养高素质人才。产出导向既符合"双碳"背景下使用英语的真实情况，又能满足大学生的学习心理需求。单元模块中设计的系列促成活动为学生搭建了完成产出任务的脚手架。这一理论融合了二语习得理论、社会文化理论与中国传统的知行合一理念，同时将育人融入用外语"做事"的实践，不但让学生获取了"双碳"通识类知识，还在潜移默化中形成正确的世界观、人生观和价值观。

三、强化价值引领，讲好中国故事

本教材将习近平生态文明思想贯穿其中，蕴含了丰富的中国元素，呈现了中国在应对全球气候变化方面所做出的承诺、为之付出的努力，以及在科技进步及发展方面取得的成就，让学生了解中国智慧、中国方案、中国力量在建设清洁美丽世界中的贡献，增强国家认同，将个人发展与社会发展、国家发展紧密相连，成为"知中国、爱中国、讲中国"的国际传播人才。

内容框架及教学建议

本教材共有八个单元，每单元对应一个主题，每个主题由四个模块组成，模块的组成理念基于对主题从初见（First Sight），到渐知（Getting to Know），到钟情（Engaging Yourself），再到相伴（Commitment）的完整认知过程。

"初见"蕴含读者与主题的第一次相遇——"初见乍惊欢"。该模块包含两个具有趣味性、知识性的视频，创造沉浸式体验，激发学生的兴趣和求知欲。该模块重点在于激发学生的学习动力，明确学习目标。

"渐知"寓意对主题的进一步探索与知晓。该模块提供了一篇与单元主题密切相关的精细阅读文章，介绍该领域的通识类知识或者最前沿的科技信息。在阅读课文之前设计了"术语解释"环节，预热相关话题，激发学生进一步探索的欲望。在课文教学环节中，教师可以重点帮助学生汲取"双碳"通识类知识和术语表达等语言知识，积累语言与内容，形成学习支架，助力学生完成单元任务。

"钟情"意味着深刻了解之后的喜爱与倾心，从而让学生探索更多与主题相关的

前言

信息。从该模块的拓展阅读文章中,学生不但可以了解世界范围内应对全球气候变暖所做出的努力和取得的成就,还能发现更多的中国元素,了解中国在"双碳"领域的政策、行动和取得的成就,培养家国情怀。课后设计了词汇和撰写课文概要的练习,强化语言积累,锻炼学生的书面表达能力,为单元产出任务打下基础。

"相伴"则代表将"双碳"相关主题的知识融入学生已有的知识体系之中,相伴终生。该部分共设计三项单元产出任务:翻译、讨论和主题展示。教师可根据任务设计不同的场景,帮助学生综合运用单元所学,结合交际技能,分步完成产出任务,培养学生表达、讲述、沟通的能力,尤其是用英语写好、讲好中国"双碳"故事的能力。

本书每单元授课时间建议4学时,教师亦可按需制订符合班级学情的教学计划和做好时间安排,关键在于充分利用本书丰富的视频和阅读材料,培养学生"双碳"领域英语学习的能动性与积极性。

本书可供高等院校非英语专业学生作为双碳通识英语、能源生态英语、能源与环境英语等通识类大学英语课程教材使用,也可作为"双碳"领域从业人员和广大英语学习爱好者拓展"双碳"领域通识类知识、提高英语综合技能的参考书。

本书得以付梓,除了编写团队的辛勤付出之外,特别感谢南京航空航天大学"十四五"规划教材建设项目、江苏省高等教育教学改革研究课题、江苏省高校"高质量公共课教学改革研究"专项课题等项目的大力资助。

由于时间及编者水平和经验有限,书中欠妥之处在所难免,期待各位专家、老师和同学在使用教材的过程中批评指正,使教材不断完善。

编者

2024年7月

Contents

Unit 1 Carbon We Live By

Phase I First Sight /2
　Video I Why Is Carbon the Key to Life? /2
　Video II Keep Up with Carbon /4

Phase II Getting to Know /7
　Warm-up Activity /7
　Active Reading: The Carbon Cycle /7

Phase III Engaging Yourself /13
　Warm-up Activity /13
　Further Reading: A Short History of the Earth's Climate /13

Phase IV Commitment /20

Vocabulary /21

Unit 2 Carbon Emissions

Phase I First Sight /28
　Video I A Simple Explanation of Climate Change /28
　Video II Europe's Climate in 2050 /30

Phase II Getting to Know /32
　Warm-up Activity /32
　Active Reading: Greenhouse Effect /32

Phase III Engaging Yourself /39
　Warm-up Activity /39
　Further Reading: Are Humans to Blame for Global Warming? /39

Phase IV Commitment /45

Vocabulary /46

Unit 3 Carbon Neutrality

Phase I   First Sight /50
　　Video I   Carbon Neutrality /50
　　Video II "Explaining Shanghai" — Carbon Reduction /52

Phase II Getting to Know /54
　　Warm-up Activity /54
　　Active Reading: Carbon Neutrality and Net Zero /54

Phase III Engaging Yourself /60
　　Warm-up Activity /60
　　Further Reading: Achieve Dual Carbon Goals in a Balanced Way /60

Phase IV Commitment /66

Vocabulary /67

Unit 4 Carbon Trading

Phase I   First Sight /72
　　Video I   What Is Carbon Trading? /72
　　Video II How Do Carbon Markets Work? /74

Phase II Getting to Know /76
　　Warm-up Activity /76
　　Active Reading: Carbon Trading — What Is It and How Does It Work? /76

Phase III Engaging Yourself /83
　　Warm-up Activity /83
　　Further Reading: China's National ETS — A Model /83

Phase IV Commitment /89

Vocabulary /90

Contents

Unit 5 Green Materials

Phase I	**First Sight**	**/94**
	Video I The Miracle Material — Graphene	/94
	Video II Materials Science for a Sustainable World	/96
Phase II	**Getting to Know**	**/99**
	Warm-up Activity	/99
	Active Reading: Materials Science Breakthroughs	/99
Phase III	**Engaging Yourself**	**/106**
	Warm-up Activity	/106
	Further Reading: What's China Doing in Materials Science?	/106
Phase IV	**Commitment**	**/113**
Vocabulary		**/114**

Unit 6 Clean Energy

Phase I	**First Sight**	**/122**
	Video I Advantages and Disadvantages of Renewable Energy	/122
	Video II Can 100 Percent Renewable Energy Power the World?	/124
Phase II	**Getting to Know**	**/127**
	Warm-up Activity	/127
	Active Reading: All We Need to Know About Clean Energy	/127
Phase III	**Engaging Yourself**	**/133**
	Warm-up Activity	/133
	Further Reading: China's Green Transition: A Feasible Way of Quitting Coal	/133
Phase IV	**Commitment**	**/141**
Vocabulary		**/142**

Unit 7 Transport Decarbonization

Phase I First Sight /148
    Video I   The Surprisingly Long History of Electric Cars /148
    Video II Can Flying Go Green? /150

Phase II Getting to Know /153
    Warm-up Activity /153
    Active Reading: A Strategy to Decarbonize the Global Transport Sector /153

Phase III Engaging Yourself /160
    Warm-up Activity /160
    Further Reading: How Can Airlines Chart a Path to Zero-Carbon Flying? /160

Phase IV Commitment /167

Vocabulary /168

Unit 8 Sustainable Living

Phase I First Sight /174
    Video I   What Is Sustainability? /174
    Video II Top Eco-Friendly Cities in the World /176

Phase II Getting to Know /179
    Warm-up Activity /179
    Active Reading: How Cities Are Going Carbon Neutral? /179

Phase III Engaging Yourself /185
    Warm-up Activity /185
    Further Reading: Green Living in China /185

Phase IV Commitment /192

Vocabulary /193

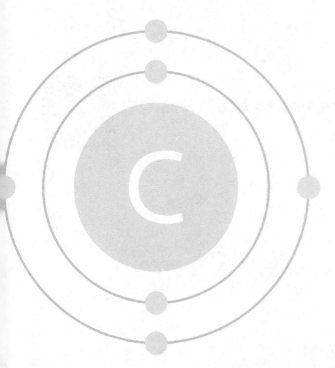

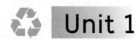

Carbon We Live By

You will die but the carbon will not; its career does not end with you. It will return to the soil, and there a plant may take it up again in time, sending it once more on a cycle of plant and animal life.

—Jacob Bronowski

Phase I First Sight

Video 1 Why Is Carbon the Key to Life?

New Words and Phrases

polymer /ˈpɒlɪmə(r)/ *n.* [高分子] 聚合物

molecule /ˈmɒlɪkjuːl/ *n.* 分子

scaffolding /ˈskæfəldɪŋ/ *n.* 脚手架

graphite /ˈɡræfaɪt/ *n.* 石墨

promiscuous /prəˈmɪskjuəs/ *adj.* 杂乱的；混杂的

chlorine /ˈklɔːriːn/ *n.* 氯气；氯

nitrogen /ˈnaɪtrədʒən/ *n.* 氮气；氮

sulfur /ˈsʌlfə(r)/ *n.* 硫；硫黄

hydrogen /ˈhaɪdrədʒən/ *n.* 氢气；氢

tetrahedral /ˌtetrəˈhiːdrəl/ *adj.* 四面体的

phosphorus /ˈfɒsfərəs/ *n.* 磷

silicon /ˈsɪlɪkən/ *n.* 硅

periodic table 元素周期表

oxidize /ˈɒksɪdaɪz/ *v.* 使氧化

stoke /stəʊk/ *v.* 煽动；激起

enzyme /ˈenzaɪm/ *n.* 酶

cytochrome /ˈsaɪtəʊˌkrəʊm/ *n.* 细胞色素

1. Watch the video and choose the best answer.

(1) Which of the following statements is **NOT** true?

    A. DNA and proteins are based on carbon.

    B. Fats and sugars are based on carbon.

    C. All the molecule cells are based on carbon.

    D. All known life forms are based on carbon.

(2) Which of the following is **NOT** the reason why carbon occupies a special place among the elements on the Earth?

    A. Carbon can form bonds with many other elements.

    B. Carbon can only form molecules of flat shape.

    C. Carbon can bond in a bunch of different ways.

    D. Carbon can form incredibly long polymers.

(3) According to the video, silicon _____.

    A. is the major building block for life in the universe

    B. is one of the two alternatives to forming life

    C. can create stable bonds with other elements

    D. shares many similarities with carbon

(4) When silicon forms a bond with oxygen, it _____.

    A. creates a solid which is known as sand

    B. creates a gas which is easy for plants to access

    C. creates a gas from which sugar is made

    D. suggests evidence to form silicon-based life

2. Watch the video again and complete the sentences with the words you hear.

(1) No known life on the Earth does not use carbon as the basic _____ of its own existence.

(2) Carbon can form less stable, more _____ bonds with elements, like oxygen.

(3) It's not inconceivable that, somewhere in the universe, elements other than carbon could be _____ blocks for life.

(4) Silicon, right below carbon on the periodic table, has popped up as a potential _____.

(5) Some life can already use silicon, though, like in these beautiful shells. And scientists are evolving _____ that can use it in other ways too.

(6) Using a technique called _____ evolution, scientists generated mutations in an enzyme called cytochrome.

(7) Over several generations, the mutated bacteria were able to produce 20 different _____ with silicon-carbon bonds.

(8) Trying to measure the _____ from a planet trillions of kilometers away, when it's sitting right next to the Sun, is a chemical trick we haven't quite figured out yet.

3. Answer the following questions according to the video.

(1) What are the similarities between carbon and silicon?

(2) Why is it hard to find alien life forms outside the solar system?

Video II Keep Up with Carbon

New Words

crust /krʌst/ *n.* 硬层；硬表面

photosynthesis /ˌfəʊtəʊˈsɪnθəsɪs/ *n.* 光合作用

nutrient /ˈnjuːtriənt/ *n.* 营养素；营养物

decay /dɪˈkeɪ/ *v.* 腐烂；腐朽

offshore /ˌɒfˈʃɔː(r)/ *adv.* 向海地；离岸地

phytoplankton /ˌfaɪtəˈplæŋktən/ *n.* 浮游植物

regulator /ˈreɡjuleɪtə(r)/ *n.* （温度）自动调节器

Unit 1 Carbon We Live By

> **gradient** /'greɪdiənt/ *n.* （温度）变化率
>
> **ecosystem** /'iːkəʊsɪstəm/ *n.* 生态系统

1. Watch the video and judge whether the following statements are TRUE or FALSE.

(1) _____ Carbon on the Earth is only stored in the ocean, the atmosphere, and the crust of the planet.

(2) _____ Plants and animals give carbon back to the soil when they die and decay.

(3) _____ The ocean holds a much greater amount of carbon than the atmosphere.

(4) _____ In the past several million years, the Earth has rarely witnessed such a high level of CO_2 in the atmosphere.

(5) _____ Climate change gives rise to warmer water and weak circulation which will affect marine life and the ecosystem.

2. Watch the video again and fill in the blanks with the words you hear.

At the ocean surface, CO_2 from the atmosphere (1) _____ the water. Tiny marine plants called phytoplankton use this CO_2 for photosynthesis. Phytoplankton are the (2) _____ of the marine food web. After animals eat the plants, they breathe out carbon or pass it up the (3) _____.

Sometimes phytoplankton die, (4) _____, and are recycling in the surface waters. Phytoplankton can also sink to the ocean floor, carrying carbon as they (5) _____. Over long time scales, this process has made the ocean floor the largest (6) _____ of carbon on the planet.

Most of the ocean's nutrients are in cold deep water. In a process called (7) _____, currents bring nutrients and carbon up to the surface. Carbon can then be released as a gas back into the atmosphere, continuing the (8) _____.

3. Answer the following questions according to the video.

(1) Why are the oceans actually a great regulator, a controller of the Earth's climate?

(2) Why will oceans become less effective at removing carbon from the atmosphere in the future?

Unit 1 Carbon We Live By

Phase II Getting to Know

Warm-up Activity

Explain the following terms according to what you've explored before class.

- carbon sink
- carbon cycle
- photosynthesis
- greenhouse gases

Active Reading

The Carbon Cycle

[1] Carbon is the backbone of life on Earth. We are made of carbon, we eat carbon, and our civilizations, our economies, our homes, and our means of transport are built on carbon.

[2] **Forged** in the heart of aging stars, carbon is the fourth most abundant element in the universe. Most of the Earth's carbon — about 65,500 billion **metric tons** — is stored in rocks. The rest is in the ocean, atmosphere, plants, soil, and fossil fuels.

[3] Carbon flows between each reservoir in an exchange are called the carbon cycle. Any change in the cycle that shifts carbon out of one reservoir puts more carbon in the other reservoirs. Changes that put carbon gases into the atmosphere result in warmer temperatures on the Earth.

[4] Over the long term, the carbon cycle seems to maintain a balance that prevents all of the Earth's carbon from entering the atmosphere or from being stored entirely in rocks. This balance helps keep the Earth's temperature relatively stable, like a **thermostat**.

[5] On very long time scales (millions to tens of millions of years), the movement of **tectonic** plates and changes in the rate at which carbon **seeps** from the Earth's

interior may change the temperature on the thermostat. The Earth has undergone such a change over the last 50 million years, from the extremely warm climates of the Cretaceous[1] to the glacial climates of the Pleistocene[2].

The Slow Carbon Cycle

[6] Through a series of chemical reactions and tectonic activity, carbon takes between 100 to 200 million years to move between rocks, soil, ocean, and atmosphere in the slow carbon cycle. On average, 10^{13} to 10^{14} grams (10 to 100 million metric tons) of carbon move through the slow carbon cycle every year. In comparison, human emissions of carbon to the atmosphere are on the order of 10^{15} grams, whereas the fast carbon cycle moves 10^{16} to 10^{17} grams of carbon per year.

[7] The movement of carbon from the atmosphere to the **lithosphere** begins with rain. Atmospheric carbon combines with water to form a weak acid — **carbonic acid** — that falls to the surface in rain. The acid **dissolves** rocks — a process called **chemical weathering** — and releases **calcium**, **magnesium**, **potassium**, or **sodium ions**. Rivers carry the ions to the ocean.

[8] In the ocean, the calcium ions combine with **bicarbonate** ions to form **calcium carbonate**, the active **ingredient** in **antacids** and the chalky white substance that dries on your **faucet** if you live in an area with hard water.

[9] In the modern ocean, most of the calcium carbonate is made by shell-building organisms (such as corals) and **plankton**. After the organisms die, they sink to the ocean floor. Over time, layers of shells and **sediment** are cemented together and turn to rock, storing the carbon in stone — **limestone** and its **derivatives**.

[10] Only 80 percent of carbon-containing rock is currently made this way. The remaining 20 percent contains carbon from living things (organic carbon) that have

---

1 Cretaceous: 白垩纪，地质年代中中生代的最后一个纪，开始于1.45亿年前，结束于6 600万年前。

2 Pleistocene: 更新世，亦称洪积世（2 588 000年前 — 11 700年前），属于地质时代第四纪的早期。

Unit 1 Carbon We Live By

been **embedded** in layers of mud. Heat and pressure compress the mud and carbon over millions of years, forming sedimentary rock such as **shale**. In special cases, when dead plant matter builds up faster than it can decay, layers of organic carbon become oil, coal, or natural gas instead of sedimentary rock like shale.

⑪ The slow cycle returns carbon to the atmosphere through volcanoes. The Earth's land and ocean surfaces sit on several moving crustal plates. When the plates collide, one sinks beneath the other, and the rock it carries melts under the extreme heat and pressure. The heated rock recombines into **silicate** minerals, releasing carbon dioxide.

⑫ When volcanoes erupt, they **vent** the gas into the atmosphere and cover the land with fresh silicate rock to begin the cycle again. At present, volcanoes emit between 130 and 380 million metric tons of carbon dioxide per year. For comparison, humans emit about 30 billion tons of carbon dioxide per year — 100–300 times more than volcanoes — by burning fossil fuels.

⑬ Chemistry regulates this dance between ocean, land, and atmosphere. If carbon dioxide rises in the atmosphere because of an increase in volcanic activity, for example, temperatures rise, leading to more rain, which dissolves more rock, creating more ions that will eventually **deposit** more carbon on the ocean floor. It takes a few hundred thousand years to rebalance the slow carbon cycle through chemical weathering.

⑭ However, the slow carbon cycle also contains a slightly faster component: the ocean. At the surface, where air meets water, carbon dioxide gas dissolves in and **ventilates** out of the ocean in a steady exchange with the atmosphere. Once in the ocean, carbon dioxide gas reacts with water molecules to release hydrogen, making the ocean more acidic. The hydrogen reacts with carbonate from rock weathering to produce bicarbonate ions.

The Fast Carbon Cycle

⑮ The time it takes carbon to move through the fast carbon cycle is measured in a lifespan. The fast carbon cycle is largely the movement of carbon through life forms on the Earth, or the **biosphere**. Between 10^{15} and 10^{17} grams (1,000 to 100,000 million metric tons) of carbon move through the fast carbon cycle every year.

[16] Carbon plays an essential role in biology because of its ability to form many bonds — up to four per atom — in a seemingly endless variety of complex organic molecules. Many organic molecules contain carbon atoms that have formed strong bonds with other carbon atoms, combining into long chains and rings. Such carbon chains and rings are the basis of living cells. For instance, DNA is made of two **intertwined** molecules built around a carbon chain.

[17] The bonds in the long carbon chains contain a lot of energy. When the chains break apart, the stored energy is released. This energy makes carbon molecules an excellent source of fuel for all living things.

[18] Plants and plankton are the main components of the fast carbon cycle. They take carbon dioxide from the atmosphere by absorbing it into their cells. Using energy from the Sun, both plants and plankton combine carbon dioxide and water to form sugar (CH_2O) and oxygen.

[19] Four things can happen to move carbon from a plant and return it to the atmosphere, but all involve the same chemical reaction. Plants break down the sugar to get the energy they need to grow. Animals (including people) eat the plants or plankton, and break down the plant sugar to get energy. Plants and plankton die and decay at the end of the growing season. Or fire consumes plants. In each case, oxygen combines with sugar to release water, carbon dioxide, and energy.

[20] In all four processes, the carbon dioxide released in the reaction usually ends up in the atmosphere. The fast carbon cycle is so tightly tied to plant life that the growing season can be seen by the way carbon dioxide **fluctuates** in the atmosphere. In the Northern Hemisphere winter, when few land plants are growing and many are decaying, atmospheric carbon dioxide concentrations climb. During the spring, when plants begin growing again, concentrations drop. It is as if the Earth is breathing.

[21] Left **unperturbed**, the fast and slow carbon cycles maintain a relatively steady concentration of carbon in the atmosphere, land, plants, and ocean. But when

Unit 1 Carbon We Live By

anything changes the amount of carbon in one reservoir, the effect ripples through the others.

(1,220 words)

1. Judge whether the following statements are TRUE or FALSE.

(1) _____ Carbon is forged in the heart of aging stars, so it is the fourth most abundant element in the universe.

(2) _____ The carbon cycle involves the exchange of carbon among different reservoirs.

(3) _____ Over the long term, the carbon cycle seems to maintain a balance that prevents most of the Earth's carbon from entering the atmosphere or from being stored entirely in rocks.

(4) _____ Through a series of chemical reactions and tectonic activities, carbon moves between rocks, soil, ocean, and atmosphere in the slow carbon cycle.

(5) _____ The fast carbon cycle refers to the movement of carbon through life forms on the Earth.

2. Complete the following sentences with the words or phrase in the box below.

| dissolved | break down | intertwine | emitting |
| regulate | undergone | unperturbed | deposited |

(1) Throughout the years, the church has _____ numerous renovations.

(2) The United States has long been the world's biggest emitter of carbon dioxide and is still the second biggest, _____ 14 percent of the global total.

(3) In a new study, researchers quantified the number of the particles _____ since the early 1930s.

(4) Its vegan formula contains vitamin C which helps _____ chlorine molecules.

(5) It is the proteins that drive and _____ critical chemical reactions throughout the human body.

(6) To create renewable paper tissue and other products, the lignin (木质素) in

wood must be cut and _____ with hazardous chemicals.

(7) The bison seemed _____ as the woman posed for the camera next to it, remaining still on the ground.

(8) As the novel unfolds, many different characters and plotlines _____.

3. Match the following technical terms with their Chinese equivalents.

I	II
(1) carbon cycle	(　) 化学风化
(2) tectonic activity	(　) 二氧化碳
(3) chemical weathering	(　) 碳循环
(4) organic molecule	(　) 光合作用
(5) biosphere	(　) 有机分子
(6) carbon dioxide	(　) 构造活动
(7) photosynthesis	(　) 浮游生物
(8) calcium carbonate	(　) 碳酸氢根离子
(9) plankton	(　) 碳酸钙
(10) bicarbonate ion	(　) 生物圈

4. Translate the following paragraphs into Chinese.

The fast carbon cycle is so tightly tied to plant life that the growing season can be seen by the way carbon dioxide fluctuates in the atmosphere. In the Northern Hemisphere winter, when few land plants are growing and many are decaying, atmospheric carbon dioxide concentrations climb. During the spring, when plants begin growing again, concentrations drop. It is as if the Earth is breathing.

Left unperturbed, the fast and slow carbon cycles maintain a relatively steady concentration of carbon in the atmosphere, land, plants, and ocean. But when anything changes the amount of carbon in one reservoir, the effect ripples through the others.

Unit 1 Carbon We Live By

Phase III Engaging Yourself

Warm-up Activity

Explain the following terms according to what you've explored before class.

- chemical weathering
- slow carbon cycle
- fast carbon cycle

Further Reading

A Short History of the Earth's Climate

[1] During the 4.6 billion years since the Earth was formed, its climate system evolves in time under the influence of its own internal dynamics and as a result of changes in external factors that affect climate, or "forcings", such as volcanic eruptions and solar variations as well as human-**induced** changes in atmospheric composition.

[2] Some 225 million years ago, the Great Dying[1] occurred when oceans absorbed vast quantities of carbon dioxide from the air, **depleting** the oxygen in the water and triggering a worldwide **bloom** of a sulfur-emitting **anaerobic** bacteria. The seas became **spiked** with acid and the air was filled with poisonous gas. The result was an event even more **cataclysmic** than the later extinction of the dinosaurs. This event is presently under active investigation given the rate at which we humans are now warming and **acidifying** the oceans.

[3] At least five major ice ages — millions to tens of millions of years, when large areas of the Earth are covered by glaciers and global temperatures are relatively

---

1 Great Dying: 大灭绝，也称为二叠纪–三叠纪灭绝事件，是显生宙"五大"大规模灭绝中规模最大的一次。

cold — have been documented over the Earth's history. The current ice age began about 3 million years ago and includes all of human history. Within a major ice age are shorter periods called "**interglacials**" (when glaciers retreat and global temperatures are warm) and "glacials" (when glaciers advance and temperatures drop). In the last million years, a major glacial occurred approximately every 100,000 years. We are now in an interglacial (warm period) that began about 11,000 years ago and includes the whole of the development of human societies and civilizations.

The Pleistocene Epoch

[4] The Pleistocene Epoch is typically defined as the period that began about 2.6 million years ago and lasted until about 11,700 years ago at the start of the current interglacial period. During this epoch the climate was **chaotic**. While parts of the globe experienced a reasonable climate, much of Eurasia and North America was buried under ice several kilometers thick. The climates across the northern continents swung from the depth of glacial **frigidity** to a relative mildness in the space of a few years. This **erratic** pattern was a feature of virtually the whole of the last 100,000 years.

The Holocene Epoch[1] and the Collapse of the Bronze Age[2]

[5] After the final **paroxysms** of the last major glacial came to an end around 12,000 years ago, the world warmed dramatically over the next two millennia. The Earth entered the Holocene Epoch and the Neolithic Era[3] began. The Earth's climate has been relatively stable for the past 10,000 years, which is widely recognized as the central reason for the explosive development of societies and civilizations.

[6] Despite relative stability, humanity in the Holocene has seen several

---

1 Holocene Epoch：全新世，第四纪的第二纪元，在更新世之后。

2 Bronze Age：青铜时代，大约从公元前 3 300 年持续到公元前 1 200 年的一段历史时期，其特点是青铜的使用。

3 Neolithic Era：新石器时代，石器时代的最后一个阶段，距今约 1 万年前至公元前 3 000 年。

significant climate-induced setbacks. From about 1,800 BCE until around 1,200 BCE, civilizations developed in the Aegean[1], Egypt, and the Near East. Commerce **flourished** in the region to the point that it can be described as the earliest known example of global trade. Then after nearly 2,000 years of growth and prosperity, a series of catastrophes caused the final collapse of the Bronze Age civilizations.

[7] A severe change in climate certainly played a role. Toward the end of the 13th BCE and the early decades of the 12th century BCE, a drought lasting about 300 years appeared to have affected the entire region. This would obviously have caused crop failures, famine, and human migrations. Scarcity of resources would have caused problems including violence at home and overseas.

[8] The collapse of the Bronze Age resulted in the destruction of almost every major city in the eastern Mediterranean world. It took four centuries before Greek society reemerged, entering what we know as the Archaic Period.

Subsequent Cooling Events

[9] There were many instances **subsequent** to the collapse of the Bronze Age where climate-induced conditions contributed to severe **setbacks**. A hundred-year cooling period in the Northern Hemisphere occurred in the 6th and 7th century CE following three immense volcanic eruptions in the western Pacific. The extreme weather events of 535 – 536 CE were the most severe and **protracted** short-term episodes of cooling in the Northern Hemisphere in the last 2,000 years. The effects were widespread, causing **unseasonal** weather, crop failure, and famine worldwide — though not nearly at the scope of what we expect in our future if the current warming goes unchecked.

[10] Another centuries-long period of global cooling started in the early 14th century and extended to the mid-19th century, causing considerable agricultural distress in Europe. The year 1816 is known as the Year Without a Summer (also the Poverty Year). Severe climate abnormalities caused average global temperatures

---

1 Aegean: 爱琴海，地中海的一部分，位于希腊半岛和小亚细亚半岛之间。

to decrease by 0.7–1.3 °F, resulting in major food shortages across the Northern Hemisphere. European peasants suffered famines, **hypothermia**, bread riots and the rise of **despotic** leaders **brutalizing** an increasingly **dispirited peasantry**.

[11] The cooling was likely caused mainly by the 1815 eruption of Mount Tambora on the island of Sumbawa[1] in present-day Indonesia, one of the most powerful volcanic eruptions in recorded history. It may also have been impacted by the Dalton Minimum[2] which ran from December 1810 to May 1823.

[12] It would be a mistake to think that humanity had no role in these climate feedback events or in determining the impact. Over-hunting, depleting agricultural practice, massive burning, tribal warring over scarce resources, scapegoating in the face of disaster — all were human responses that **exacerbated** the outside forces and contributed to the outcomes.

[13] Externally-caused factors continue to pose threats; for example, a dozen or more earthquake storms of varying **magnitude** have been identified in the 21st century. But clearly, the greatest threats now to the stability of the Earth's life-supporting climate are those we humans are creating — at an unprecedented and truly alarming rate.

[14] It is also a mistake to think that all will be fine because the Earth has had major climate shifts before. Recent research shows that no previous climate event over the past 2,000 years was even remotely equivalent in degree or extent to the warming over the past few decades.

[15] The planet and climate will survive in one form or another, but humans will not. Fortunately, for the first time in history, we are in a position to know what's happening, the impact we are having, and what will happen if we do nothing.

(1,037 words)

---

1 Sumbawa：松巴哇，印度尼西亚西努沙登加拉省岛屿，小巽他群岛之一。

2 Dalton Minimum：道尔顿极小期，太阳黑子数量低的一个时期，以英国气象学家约翰·道尔顿的名字命名。

Unit 1　Carbon We Live By

1. the Near East 近东

政治地理术语，相对中东、远东地区而言的概念，指距离西欧较近的国家和地区。过去主要指欧洲的巴尔干国家、亚洲的地中海沿岸国家和东地中海岛国塞浦路斯；第二次世界大战后，此称已被"中东"取代，但两者常通用。"近东"一般用在文明史上，而"中东"常用在政治上。

2. Year Without a Summer 无夏之年

指1816年。受1815年印度尼西亚坦博拉火山爆发的影响，北半球天气出现严重反常，欧洲、北美洲及亚洲都出现灾情，夏天出现罕见低温；欧洲及美洲农业生产受影响尤甚。

3. Mount Tambora 坦博拉火山

荷属东印度群岛（如今的印度尼西亚）中的海岛。坦博拉火山的爆发令数万人丧生，浓重的火山灰令庄稼颗粒无收，不少人更是死于其后的饥荒。

4. Archaic Period 古风时期

古风的字面意思是远古和古朴，指黑暗时代之后古典时代之前的希腊历史，这个时期从公元前8世纪持续到公元前5世纪初。古风时期是希腊城邦制度迅速发展、希腊人不断向本土之外殖民扩张的时代，其艺术先后呈现出几何风格、东方风格和古风风格特征。

1. Read the text and choose the best answer.

(1) Compared with the Great Dying, the extinction of the dinosaurs _____.

　　A. occurred when oceans absorbed greater quantities of carbon dioxide

　　B. occurred due to a worldwide bloom of a sulfur-emitting anaerobic bacteria

　　C. has drawn much more attention of the public nowadays

　　D. was believed to be less cataclysmic

(2) In Paragraph 3, we learn that _____.

   A. ice ages have been documented only five times

   B. interglacials and glacials are only included in the current ice age

   C. a major glacial occurred around every 100,000 years in every ice age

   D. the current ice age covers all of human history

(3) Which of the following statements is true?

   A. The climate during the Pleistocene Epoch was chaotic.

   B. The climate during the Holocene Epoch was hot.

   C. The Pleistocene Epoch began at the start of the glacial period.

   D. The Holocene Epoch was followed by the Neolithic Era.

(4) The root reason for the collapse of the Bronze Age civilizations is _____.

   A. a severe change in climate

   B. scarcity of resources

   C. a drought lasting about 300 years

   D. violence at home and overseas

(5) Which of the following statements is true?

   A. Previous climate events over the past 2,000 years were equivalent to the current global warming.

   B. The greatest threats now to the Earth's life-supporting climate are created by humans.

   C. The planet and climate as well as humans will survive in one form or another.

   D. External factors pose the greatest threats to climate shifts.

2. **Replace the underlined parts with the words in the box below.**

| spiked | flourish | protracted | exacerbated |
| paroxysm | subsequent | scarcity | depleted |

(1) The oxygen in the water was <u>reduced dramatically</u> due to vast quantities of carbon dioxide in the oceans.

Unit 1   Carbon We Live By

(2) New telephone orders have <u>increased considerably</u> in the last two years.

(3) After the final <u>outburst</u> of the last major glacial came to an end, the world warmed dramatically over the next two millennia.

(4) No village along the railroad failed to <u>prosper</u>.

(5) The <u>shortage</u> of medical supplies was becoming critical.

(6) The first meeting will be in the City Hall, but all <u>following</u> meetings in the school.

(7) It was going to be another <u>prolonged</u> day; he had to stay alert and miss nothing.

(8) High fuel costs certainly <u>worsened</u> Indian airlines' woes.

3. Write a summary of the text in 120–150 words. You may use the given words or expressions in the box.

feedback forcings	the Great Dying
ice ages	the Pleistocene Epoch
chaotic climate	the Holocene Epoch
climate-induced setbacks	the collapse of the Bronze Age
cooling events	humanity

Phase IV Commitment

1. Translate the following paragraph into English.

中华文明历来崇尚天人合一、道法自然。但人类进入工业文明时代以来，在创造巨大物质财富的同时，人与自然深层次矛盾日益凸显。大自然孕育抚养了人类，人类应该以自然为根，尊重自然、顺应自然、保护自然。中国站在对人类文明负责的高度，积极应对气候变化，构建人与自然生命共同体，推动形成人与自然和谐共生新格局。

2. Work in pairs to discuss the following topics.

The carbon cycle is the process by which carbon moves between different reservoirs. Over the long term, the carbon cycle seems to maintain a balance that helps keep the Earth's temperature relatively stable.

(1) What impacts have human activities exerted on the carbon cycle?
(2) How can human impacts on the carbon cycle be prevented?

3. Work in groups to make a presentation on any topic related to the theme of the unit.

Vocabulary

Active Reading: The Carbon Cycle

forge	/fɔːdʒ/	v.	to shape metal by heating it in a fire and hitting it with a hammer; to make an object in this way	锻造；铸造
metric ton	/ˌmetrɪk 'tʌn/	n.	a unit for measuring weight, equal to 1,000 kilograms	公吨
thermostat	/'θɜːməstæt/	n.	an instrument used for keeping a room or a machine at a particular temperature	恒温器；温度自动调节器
tectonic	/tek'tɒnɪk/	adj.	(geology) relating to the structure of the Earth's surface	地壳构造的
seep	/siːp/	v.	to flow slowly through small holes or spaces	渗；渗透
lithosphere	/'lɪθəsfɪə(r)/	n.	(geology) the layer of rock that forms the outer part of the Earth	岩石圈；岩石层
dissolve	/dɪ'zɒlv/	v.	(of a solid) to mix with a liquid and become part of it	溶；（使）溶解
calcium	/'kælsiəm/	n.	a silver-white metal that helps to form teeth, bones, and chalk	（化学元素）钙
magnesium	/mæg'niːziəm/	n.	a common silver-white metal that burns with a bright white flame	（化学元素）镁
potassium	/pə'tæsiəm/	n.	a common soft silver-white metal that usually exists in combination with other substances, used for example, in farming	（化学元素）钾

sodium	/ˈsəʊdiəm/	*n.*	a common silver-white metal that usually exists in combination with other substances, for example, in salt	（化学元素）钠
ion	/ˈaɪən/	*n.*	(technical) an atom which has been given a positive or negative force by adding or taking away an electron	离子
bicarbonate	/ˌbaɪˈkɑːbənət/	*n.*	(chemistry) a salt made from carbonic acid containing carbon, hydrogen, and oxygen together with another element	碳酸氢盐
ingredient	/ɪnˈɡriːdiənt/	*n.*	one of the things from which sth. is made, especially one of the foods that are used together to make a particular dish	成分；（尤指烹饪）原料
antacid	/æntˈæsɪd/	*n.*	a substance that gets rid of the burning feeling in your stomach when you have eaten too much, drunk too much alcohol, etc.	解酸药；抗酸药
faucet	/ˈfɔːsɪt/	*n.*	the thing that you turn on and off to control the flow of water from a pipe	水龙头；旋塞
plankton	/ˈplæŋktən/	*n.*	the very small forms of plant and animal life that live in water	浮游生物
sediment	/ˈsedɪmənt/	*n.*	(geology) solid substances that settle at the bottom of a liquid	沉淀物；沉积物

limestone	/ˈlaɪmstəʊn/	n.	a type of white rock that contains calcium, used in building and in making cement	石灰岩
derivative	/dɪˈrɪvətɪv/	n.	sth. that has developed or been produced from sth. else	派生物；衍生物
embed	/ɪmˈbed/	v.	to put sth. firmly and deeply into sth. else, or to be put into sth. in this way	把……牢牢地嵌入
shale	/ʃeɪl/	n.	a smooth soft rock which breaks easily into thin flat pieces	页岩；泥板
silicate	/ˈsɪlɪkeɪt/	n.	a compound of silica which does not dissolve	（矿物）硅酸盐
vent	/vent/	v.	to express feelings, especially anger, strongly	表达；发泄（感情）
deposit	/dɪˈpɒzɪt/	v.	to leave a layer of sth. on the surface of sth., especially gradually and over a period of time	使沉积；使沉淀
ventilate	/ˈventɪleɪt/	v.	to allow fresh air to enter and move around a room, building, etc.	通风；使通气
biosphere	/ˈbaɪəsfɪə(r)/	n.	(technical) the part of the world in which animals, plants, etc. can live	生物圈
intertwine	/ˌɪntəˈtwaɪn/	v.	to be or become very closely connected with sth. or sb. else	缠绕；纠缠
fluctuate	/ˈflʌktʃueɪt/	v.	to keep changing and becoming higher and lower	起伏不定
unperturbed	/ˌʌnpəˈtɜːbd/	adj.	not worried or annoyed by sth. that has happened	不担忧的；平静的；镇定的

carbonic acid				碳酸
chemical weathering				化学风化
calcium carbonate				碳酸钙

Further Reading: A Short History of the Earth's Climate

induce	/ɪnˈdjuːs/	v.	to cause sth.	引起；导致
deplete	/dɪˈpliːt/	v.	to reduce the amount of sth. that is present or available	大量减少；耗尽
bloom	/bluːm/	n.	the state or period of greatest beauty, freshness, or vigor	最美的状态；全盛期
anaerobic	/ˌænəˈrəʊbɪk/	adj.	not needing oxygen in order to live	厌氧的
spike	/spaɪk/	v.	to increase quickly and by a large amount	（浓度）激增；突增
cataclysmic	/ˌkætəˈklɪzmɪk/	adj.	(of a natural event) causing sudden and violent change	天灾的；自然剧变的
acidify	/əˈsɪdɪfaɪ/	v.	to become or make sth. become an acid	（使）变成酸；酸化
interglacial	/ˌɪntəˈɡleɪsɪəl/	n.	also known as an interglacial period, a geological interval of warmer global average temperature that separates consecutive glacial periods within an ice age	间冰期（尤指更新世）
chaotic	/keɪˈɒtɪk/	adj.	without any order; in a completely confused state	混乱的；无秩序的

Unit 1 Carbon We Live By

frigidity	/frɪˈdʒɪdəti/	*n.*	the state of being very cold	寒冷
erratic	/ɪˈrætɪk/	*adj.*	not happening at regular times; not following any plan or regular pattern; that you cannot rely on	不稳定的；难以预测的
paroxysm	/ˈpærəksɪzəm/	*n.*	a sudden, short attack of pain, coughing, shaking, etc.	（疾病周期性）发作；突发
flourish	/ˈflʌrɪʃ/	*v.*	to develop quickly and be successful or common	繁荣；昌盛；兴旺
subsequent	/ˈsʌbsɪkwənt/	*adj.*	happening or coming after sth. else	随后的；紧接的
setback	/ˈsetbæk/	*n.*	a difficulty or problem that delays or prevents sth., or makes a situation worse	挫折；阻碍
protracted	/prəˈtræktɪd/	*adj.*	used to describe sth. that continues for a long time, especially if it takes longer than usual, necessary, or expected	延长的；拖延的
unseasonal	/ʌnˈsiːzənəl/	*adj.*	not typical of or not suitable for the time of year	无季节特征的；不合节令的
hypothermia	/ˌhaɪpəʊˈθɜːmiə/	*n.*	a serious medical condition caused by extreme cold	体温过低；低温症
despotic	/dɪˈspɒtɪk/	*adj.*	connected with or typical of a leader with great power, especially one who uses it in a cruel way	暴虐的；暴君的；专政的
brutalize	/ˈbruːtəlaɪz/	*v.*	to make sb. unable to feel normal human emotions such as pity	残酷地对待；使像野兽般残忍；使变得残酷无情

dispirited	/dɪˈspɪrɪtɪd/	*adj.*	having no hope or enthusiasm	气馁的；垂头丧气的；心灰意懒的
peasantry	/ˈpezəntri/	*n.*	all the peasants in a region or country	农民（总称）
exacerbate	/ɪgˈzæsəbeɪt/	*v.*	to make a bad situation worse	使恶化；使加剧
magnitude	/ˈmægnɪtjuːd/	*n.*	the great size or importance of sth.; the degree to which sth. is large or important	巨大；重大；重要性

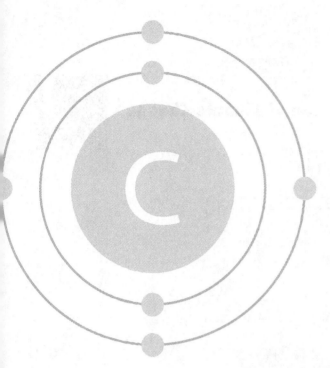
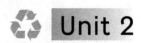

Unit 2

Carbon Emissions

Climate change is the environmental challenge of this generation, and it is imperative that we act before it's too late.

—*Anonymous*

Phase I First Sight

Video 1 A Simple Explanation of Climate Change

New Words and Phrases

Fahrenheit /'færənhaɪt/ *n.* 华氏度

polar ice cap 极地冰冠

Antarctic ice sheet 南极冰原

metropolis /mə'trɒpəlɪs/ *n.* 大都会；大城市

destabilize /ˌdiː'steɪbəlaɪz/ *v.* 使动摇

refugee /ˌrefju'dʒiː/ *n.* 避难者；逃亡者；难民

trigger /'trɪɡə(r)/ *v.* 引起

chaos /'keɪɒs/ *n.* 混乱；杂乱

insulation /ˌɪnsju'leɪʃən/ *n.* 隔热材料

1. Watch the video and choose the best answer.

(1) Which of the following statements is **NOT** true?

   A. Climate disasters will destabilize entire nations.

   B. Climate disasters will send millions of refugees across borders.

   C. Climate disasters will cause widespread extinction of all species on the Earth.

   D. Climate disasters will melt polar ice caps.

(2) If carbon emissions continue to rise unchecked, the only thing in our control is to _____.

   A. expect the Earth to turn out to be less sensitive to greenhouse gases

   B. make plants and animals adapt quickly

Unit 2　Carbon Emissions

　　C. hope technological breakthroughs help society adjust to climate change

　　D. limit emissions using all available tools and best behaviors

(3) If oceans rise, we will have to _____.

　　A. abandon all our coastal cities

　　B. adapt to an altered coastline sooner or later

　　C. stop the ice in the poles from melting completely

　　D. keep the emissions at the current level

(4) If all the ice in the poles melted, there would be a total rise of sea levels between 80 to _____ feet.

　　A. 120

　　B. 140

　　C. 160

　　D. 180

(5) Studies of past climate conditions indicate the following results **EXCEPT** that _____.

　　A. the Earth warms up

　　B. the ice in the poles melts

　　C. oceans rise

　　D. animals adapt quickly

2. Watch the video again and complete the sentences with the words you hear.

(1) The average temperature on the _____ of the planet has already increased 1.7 degrees Fahrenheit since 1880.

(2) The sudden collapse of agriculture would _____ immediate chaos in society.

(3) The melting of the Antarctic ice _____ would lead to rapidly rising seas.

(4) Humans are pumping carbon _____ into the air far faster than nature ever has before us.

(5) Scientists have been publishing strong evidence that warming is making _____ and heat waves more frequent.

(6) As more resources are devoted to solving the problem, our chances at big technological _____ are improving.

(7) You can reduce your carbon _____ by doing things like plugging leaks in your home insulation.

(8) Scientists only have the Earth's history to base their _____ on, which suggests that the rate has occasionally hit 1 foot per decade.

3. Answer the following questions according to the video.

(1) What will happen if emissions continue to rise unchecked in the longer term?

(2) What can we do to reduce carbon footprint in our daily lives?

Video II Europe's Climate in 2050

New Words and Phrases

mitigate /ˈmɪtɪɡeɪt/ *v.* 减轻；缓和

Mediterranean /ˌmedɪtəˈreɪniən/ *adj.* 地中海的

Mediterranean basin 地中海盆地

scorching /ˈskɔːtʃɪŋ/ *adj.* 酷热的

arctic /ˈɑːktɪk/ *adj.* 北极的

1. Watch the video and judge whether the following statements are TRUE or FALSE.

(1) _____ An unprecedented acceleration in climate change will most probably continue at least until 2030.

(2) _____ The current significant increase in the Earth's temperature is a direct consequence of greenhouse gases.

(3) _____ The Paris Agreement aims to mitigate the effects of climate change and stabilize the increase in global warming below 4 ℃.

Unit 2 Carbon Emissions

(4) _____ As a global phenomenon, the impact of climate change will be similar across different regions of the world.

(5) _____ Climate change occurs more rapidly in mountains than in the surrounding plains.

2. Watch the video again and fill in the blanks with the words you hear.

The warming in (1) _____ Europe is higher than the global average. This means, for example, that a typical summer of 2050 will be hot and even scorching for the (2) _____ of the Mediterranean (3) _____, because in summer much stronger warming is expected in southern Europe than anywhere else in Europe, at around one degree higher than the global average.

In southern regions, the decrease in (4) _____ and the increase in evaporation will (5) _____ the progressive drying-out of the soil. On a positive note, the increase in surface solar radiation could provide a boost for (6) _____ energy production.

Even though the overall Mediterranean climate will become (7) _____, the region could also see some much heavier autumnal rainfall. All of Europe will be subject to longer and more exceptional heat (8) _____ and episodes like the one in 2003 will become increasingly common by 2050.

3. Answer the following questions according to the video.

(1) What will happen in northeastern Europe in winter due to the effects of global warming?

(2) What would be the possible impacts of climate change on rainfall over the European continent?

31

Phase II Getting to Know

Warm-up Activity

Explain the following terms according to what you've explored before class.

- carbon footprint
- greenhouse effect
- global community
- ice age

Active Reading

Greenhouse Effect

[1] Global warming describes the current rise in the average temperature of the Earth's air and oceans. Global warming is often described as the most immediate example of climate change.

[2] The Earth's climate has changed many times. Our planet has gone through multiple ice ages, in which **ice sheets** and glaciers covered large portions of the Earth. It has also gone through warm periods when temperatures were higher than they are today.

[3] Past changes in the Earth's temperature happened very slowly, over hundreds of thousands of years. However, the recent warming trend is happening much faster than it ever has. Natural cycles of warming and cooling are not enough to explain the amount of warming we have experienced in such a short time — only human activities can account for it. Scientists worry that the climate is changing faster than some living things can adapt to it.

[4] In 1988, the World Meteorological Organization and the United Nations Environment Program established a committee of **climatologists, meteorologists,** geographers, and other scientists from around the world. This Intergovernmental Panel

on Climate Change (IPCC) includes thousands of scientists who review the most up-to-date research available related to global warming and climate change. The IPCC evaluates the risk of climate change caused by human activities.

[5] According to the IPCC's most recent report (in 2007), the Earth's average surface temperatures have risen about 0.74 °C during the past 100 years. The increase is greater in northern **latitudes**. The IPCC also found that land regions are warming faster than oceans. The IPCC states that most of the temperature increase since the mid-20th century is likely due to human activities.

[6] Human activities contribute to global warming by increasing the greenhouse effect. The greenhouse effect happens when certain gases — known as greenhouse gases — collect in the Earth's atmosphere. These gases, which occur naturally in the atmosphere, include carbon dioxide, **methane**, **nitrogen oxide**, and fluorinated gases sometimes known as **chlorofluorocarbons** (CFCs).

[7] Greenhouse gases let the Sun's light shine onto the Earth's surface, but they trap the heat that reflects up into the atmosphere. In this way, they act like the insulating glass walls of a greenhouse. The greenhouse effect keeps the Earth's climate comfortable. Without it, surface temperatures would be cooler by about 33 °C, and many life forms would freeze.

[8] Since the Industrial Revolution in the late 1700s and early 1800s, people have been releasing large quantities of greenhouse gases into the atmosphere. That amount has skyrocketed in the past century. **Greenhouse gas emissions** increased 70 percent between 1970 and 2004. Emissions of carbon dioxide, the most important greenhouse gas, rose by about 80 percent during that time. The amount of carbon dioxide in the atmosphere today far **exceeds** the natural range seen over the last 650,000 years.

[9] Most of the carbon dioxide that people put into the atmosphere comes from burning fossil fuels such as oil, coal, and natural gas. Cars, trucks, trains, and planes all burn fossil fuels. Many electric power plants also burn fossil fuels.

Effects of Global Warming

⑩ Even slight rises in average global temperatures can have huge effects. Perhaps the biggest, most obvious effect is that glaciers and ice caps melt faster than usual. The meltwater drains into the oceans, causing sea levels to rise and oceans to become less salty.

⑪ Ice sheets and glaciers advance and retreat naturally. As the Earth's temperature has changed, the ice sheets have grown and shrunk, and sea levels have fallen and risen. Ancient corals found on land in Florida, Bermuda, and the Bahamas show that the sea level must have been five to six meters higher 130,000 years ago than it is today. The Earth doesn't need to become oven-hot to melt the glaciers. Northern summers were just 3–5 °C warmer during the time of those ancient fossils than they are today.

⑫ However, the speed at which global warming is taking place is **unprecedented**. The effects are unknown.

⑬ Glaciers and ice caps cover about 10 percent of the world's landmass today. They hold about 75 percent of the world's fresh water. If all of this ice melted, sea levels would rise by about 70 meters. The IPCC reported that the global sea level rose about 1.8 millimeters per year from 1961 to 1993, and 3.1 millimeters per year since 1993.

⑭ Rising sea levels could flood coastal communities, **displacing** millions of people in areas such as Bangladesh, the Netherlands, and the U.S. state of Florida. Forced migration would impact not only those areas, but the regions to which the "climate refugees" flee. Millions more people in countries like Bolivia, Peru, and India depend on glacial meltwater for drinking, irrigation, and **hydroelectric power**. Rapid loss of these glaciers would devastate those countries.

⑮ Glacial melt has already raised the global sea level slightly. However, scientists are discovering ways the sea level could increase even faster. For example, the melting of the Chacaltaya Glacier in Bolivia has exposed dark rocks beneath it. The rocks absorb heat from the Sun, speeding up the melting process.

[16] Many scientists use the term "climate change" instead of "global warming". This is because greenhouse gas emissions affect more than just temperature. Another effect involves changes in **precipitation** like rain and snow. Patterns in precipitation may change or become more extreme. Over the course of the 20th century, precipitation increased in eastern parts of North and South America, northern Europe, and northern and central Asia. However, it has decreased in parts of Africa, the Mediterranean, and parts of southern Asia.

Future Change

[17] Nobody can look into a crystal ball and predict the future with certainty. However, scientists can make estimates about future population growth, greenhouse gas emissions, and other factors that affect climate. They can enter those estimates into computer models to find out the most likely effects of global warming.

[18] The IPCC predicts that greenhouse gas emissions will continue to increase over the next few decades. As a result, they predict the average global temperature will increase by about 0.2 °C per decade. Even if we reduce greenhouse gas and aerosol emissions to their 2,000 levels, we can still expect a warming of about 0.1 °C per decade.

[19] IPCC data also suggest that the frequency of heat waves and extreme precipitation will increase. Weather patterns such as storms and **tropical cyclones** will become more intense. Storms themselves may be stronger, more frequent, and longer-lasting. They would be followed by stronger **storm surges**, the immediate rise in sea level following storms. Storm surges are particularly damaging to coastal areas because their effects (flooding, erosion, damage to buildings and crops) are lasting.

(1,111 words)

Notes

1. World Meteorological Organization (WMO) 世界气象组织

联合国的专门机构之一，致力于研究地球大气的现状和特性，以及与海洋的相互作用。它也是关于气候及由此形成的水资源的分布方面的权威机构。

2. Intergovernmental Panel on Climate Change (IPCC) 联合国政府间气候变化专门委员会

世界气象组织及联合国环境规划署于1988年联合建立的政府间机构，其主要任务是对气候变化科学知识的现状、气候变化对社会和经济的潜在影响、如何适应和减缓气候变化的对策等进行评估。

1. Judge whether the following statements are TRUE or FALSE.

(1) _____ Global warming is often believed to be the most important cause of climate change.

(2) _____ The temperature on our planet has kept increasing through multiple ice ages, and now it has reached the highest point in human history.

(3) _____ IPCC is a committee of scientists which assesses the risk of climate change caused by human activities.

(4) _____ Greenhouse gases act like the insulating glass walls of a greenhouse, which trap the heat from the Sun that reflects up into the atmosphere.

(5) _____ Emissions of carbon dioxide rose by about 80 percent from 1970 to 2004.

2. Complete the following sentences with the words or phrases in the box below.

estimates	account for	ice sheet	tropical
displaced	unprecedented	drain into	precipitation

Unit 2 Carbon Emissions

(1) There has been a(n) _____ demand for second-hand furniture.

(2) Agricultural fertilizers often _____ aquatic ecosystems and spur a frenzy of growth.

(3) No doubt the _____ preserves specimens that would weather away more quickly in other regions.

(4) It is estimated that 500,000 refugees have been _____ by the Civil War.

(5) How do you _____ the company's alarmingly high staff turnover?

(6) The glacial system is governed by two basic climatic variables: _____ and temperature.

(7) Quotes and _____ can be prepared by a computer on the spot.

(8) When the land is exposed to the harsh _____ sun and torrential rain, it quickly becomes infertile.

3. Match the following technical terms with their Chinese equivalents.

I	II
(1) greenhouse effect	( ) 太阳磁场活动
(2) water vapor	( ) 水力发电
(3) ice cap	( ) 冰原
(4) greenhouse gas emission	( ) 降水量
(5) fossil fuel	( ) 冰川
(6) solar magnetic activity	( ) 水蒸气
(7) glacier	( ) 温室效应
(8) ice sheet	( ) 化石燃料
(9) hydroelectric power	( ) 冰冠
(10) precipitation	( ) 温室气体排放

4. Translate the following paragraphs into Chinese.

Many scientists use the term "climate change" instead of "global warming". This is because greenhouse gas emissions affect more than just temperature. Another effect involves changes in precipitation like rain and snow.

IPCC data suggest that the frequency of heat waves and extreme precipitation will increase. Weather patterns such as storms and tropical cyclones will become more intense. Storms themselves may be stronger, more frequent, and longer-lasting. They would be followed by stronger storm surges, the immediate rise in sea level following storms. Storm surges are particularly damaging to coastal areas because their effects (flooding, erosion, damage to buildings and crops) are lasting.

Phase III Engaging Yourself

Warm-up Activity

Explain the following terms according to what you've explored before class.

- IPCC
- Kyoto Protocol
- fossil fuel

Further Reading

Are Humans to Blame for Global Warming?

[1] Global warming is an environmental, economic, scientific, and political problem of the first order, and one doubly difficult to address because its dangers lie decades in the future. So if we are to act now to head it off, we must first **scrutinize** what is known about the nature of the threat. Should we place our faith in the Kyoto Protocol, which sets firm limits on human emissions of greenhouse gases?

[2] Circumstantial evidence does indeed point to our **profligate** burning of **fossil fuels** and perhaps also to its impact on global warming. Since 1900 the global temperature of the Earth's atmosphere and ocean surface waters has risen by 0.5−1 °C, and the prime suspect is atmospheric CO_2, which is second only to **water vapor** in its greenhouse effect. Since 1860, when the Industrial Revolution and soaring population growth led to widespread consumption of fossil fuels, the volume of atmospheric CO_2 has increased by about 28 percent.

[3] The increase began slowly, rising from 290 parts per million (ppm) in 1860 to 295 ppm in 1900. But it then accelerated rapidly, reaching 310 ppm in 1950 and 370 ppm in 2000, with half of the total gain of 80 ppm occurring just since 1975. **Numerical** global climate models suggest that a doubling of the current atmospheric accumulation of CO_2 will produce further warming of 3−5 °C, perhaps as soon as 2050.

[4] The consequences of this would be devastating: Inland areas **desiccated**, low-lying coastal regions battered and flooded as polar ice melts and sea levels rise, and possibly further warming and a runaway greenhouse effect due to an increase in atmospheric water vapor. The only **rational** course of action would seem to be to curtail global consumption of fossil fuels, as the Kyoto Protocol's proponents contend, and invest in alternative energy sources.

[5] While researchers created impressive global climate models in recent years, they are the first to admit that such models can include only a fraction of the many physical forces that together determine the climate and global mean temperature. For example, studies during the past 20 years have shown that changes in solar magnetic activity cause the Sun's brightness to vary by 0.1 percent, and that the average annual temperature in the northern Temperate Zone has tracked the level of solar activity over the last 1,000 years.

[6] Indeed, monitoring of other solar-type stars has revealed one whose brightness decreased by 0.5 percent in a period of five years, during which its magnetic activity declined sharply, suggesting that the Sun behaves similarly. Core samples from the Greenland ice cap, for example, show occasional sudden drops in temperature. Contrary to what climate models focusing on greenhouse gas emissions would predict, the samples show that a decline in atmospheric CO_2 followed, rather than preceded, these frigid intervals.

[7] What, then, is responsible for global warming so far? A safe bet is that from 1900 until 1950, global warming was driven mainly by the solar brightening, as solar magnetic activity increased by a factor of two or three during this period. Atmospheric CO_2 could not have been a major contributor, because it had increased by only about 7 percent before 1950, when the warming leveled off for a couple of decades. After 1950, however, solar activity showed no significant rise, while atmospheric CO_2 increased by 20 percent, accounting for the warming from 1970 to 2000. Atmospheric CO_2 is therefore presumably the controlling factor for the coming century as well.

Unit 2 Carbon Emissions

⁸ But this does not mean that human emissions are responsible for the growing accumulation of atmospheric CO_2. The atmosphere contains about 750 **gigatons** (Gt) of CO_2, while total annual human emission is approximately 5.5 Gt, thus adding annually roughly 0.7 percent of the total. However, there is also an estimated exchange of 90 Gt per year between the atmosphere and the oceans. This means that human CO_2 emissions do not simply linger and accumulate in the atmosphere. They are rapidly distributed to the ocean surface, so that atmospheric CO_2 remains at an **equilibrium** level.

⁹ This equilibrium is, in turn, determined by the temperature of ocean surface water. So it is plausible that the solar-driven ocean warming between 1900 and 1950 started things off by shifting the equilibrium toward higher concentrations of CO_2 in the atmosphere, accelerating global warming since then. Therefore, human CO_2 emissions may not be the solo determining factor of global warming.

¹⁰ Yet the threat posed by global warming is nonetheless real, and focusing on human CO_2 emissions is not necessarily "bad science", just incomplete science. For example, aside from solar magnetic activity, the Sun affects the Earth's climate in several other ways, including ultraviolet warming of the **upper stratosphere**, the nucleation of aerosols, and cloud formation. The climate is also subject to the rate of water vapor exchange between the atmosphere and the Earth's surface, which requires taking into account ocean currents, wind, and geography.

¹¹ All of these contributing effects must be understood quantitatively in order to produce an accurate model of global climate change and we remain far from that point. So the only rational response is to research aggressively into the many unknown factors: the physics of cloud formation, the dynamic coupling of the upper stratosphere to the lower atmosphere, and the accumulation of atmospheric water vapor. If effective solutions are to be found, they must await a fuller definition of the problem.

(895 words)

Note

Kyoto Protocol《京都议定书》

全称《联合国气候变化框架公约的京都议定书》，是《联合国气候变化框架公约》(United Nations Framework Convention on Climate Change, UNFCCC) 的补充条款。议定书于1997年12月11日在日本东京都通过，2005年2月16日生效。截至2023年7月，《联合国气候变化框架公约》共有198个缔约方，《京都议定书》共有192个缔约方。中国于1998年5月签署了《京都议定书》。

1. Read the text and choose the best answer.

(1) In Paragraph 2, we learn that _____.

　A. global warming only results from the burning of fossil fuels

　B. the Industrial Revolution and increasing population growth contributed to the widespread consumption of fossil fuels

　C. water vapor is second to atmospheric CO_2 in its greenhouse effect

　D. since 1900, the volume of atmospheric CO_2 has increased by about 28 percent

(2) Global warming may lead to the following possible consequences **EXCEPT** _____.

　A. drier inland areas

　B. battered and flooded low-lying coastal regions

　C. a runaway greenhouse effect

　D. a decrease in atmospheric water vapor

(3) Which of the following statements is true?

　A. Researchers have created global climate models that include all factors determining the global temperature.

　B. Studies show that changes in solar magnetic activity have caused the Sun's brightness to vary by 0.2 percent.

　C. The average annual temperature in the northern Temperate Zone has tracked the solar activity level over the last 1,000 years.

D. The average annual temperature in the southern Temperate Zone has tracked the solar activity level over the last 1,000 years.

(4) According to the text, atmospheric CO_2 had increased by _____ percent before 1950.

A. 7

B. 9

C. 20

D. 25

(5) Which of the following is **NOT** the way the Sun affects the Earth's climate?

A. Solar magnetic activity.

B. Ultraviolet warming of the upper stratosphere.

C. Water vapor exchange.

D. Cloud formation.

2. **Replace the underlined parts with the words or phrases in the box below.**

| lingered | aside from | driven | accurate |
| subject to | revealed | rational | controlling |

(1) He asked you to look at both sides of the case and come to a <u>sensible</u> decision.

(2) Details of the murder were <u>disclosed</u> by the local paper.

(3) If you love what you do, you will be <u>motivated</u> by the desire to achieve success.

(4) The low environmental temperature is still the primary <u>determining</u> factor.

(5) The faint fragrance of her perfume <u>stayed</u> in the room.

(6) <u>Apart from</u> technical challenges, the project will also pose a severe financial challenge.

(7) Their participation is <u>influenced by</u> a number of important provisos.

(8) Police have stressed that this is the most <u>precise</u> description of the killer to date.

3. Write a summary of the text in 120–150 words. You may use the given words or expressions in the box.

> fossil fuels carbon dioxide greenhouse effect
> equilibrium global temperature water vapor
> human carbon dioxide emissions unknown factors

Unit 2　Carbon Emissions

Phase IV　Commitment

1. Translate the following paragraph into English.

　　气候变化是全人类面临的共同挑战。地球是人类唯一赖以生存的家园，面对全球气候挑战，人类是一荣俱荣、一损俱损的命运共同体，没有哪个国家能独善其身。世界各国应该加强团结、推进合作，携手共建人类命运共同体。中国作为负责任的国家，愿与国际社会一起共同努力、并肩前行，助力《巴黎协定》行稳致远，为全球应对气候变化做出更大贡献。

2. Work in pairs to discuss the following topics.

(1) Global warming, an environmental, economic, scientific, and political problem to mankind, is the most immediate example of climate change. Talk about possible contributors to global warming and their impacts on the Earth.

(2) Confronted with the challenges of climate change, we should limit carbon emissions to build a global community of shared future together. Talk about what future efforts we can make to reduce carbon footprint in our daily lives.

3. Work in groups to make a presentation on any topic related to the theme of the unit.

Vocabulary

Active Reading: Greenhouse Effect

climatologist	/ˌklaɪməˈtɒlədʒɪst/	*n.*	an expert in the scientific study of climate	气候学家
meteorologist	/ˌmiːtiəˈrɒlədʒɪst/	*n.*	a scientist who studies the Earth's atmosphere and its changes, especially predicting what the weather will be like	气象学家
latitude	/ˈlætɪtjuːd/	*n.*	the distance of a place north or south of the equator	纬度
methane	/ˈmiːθeɪn/	*n.*	a gas that you cannot see or smell, which can be burned to give heat	甲烷
chlorofluorocarbon (CFC)	/ˌklɔːrəʊ-fluərəʊˈkɑːbən/	*n.*	a compound containing carbon, fluorine, and chlorine that is harmful to the ozone layer	含氯氟烃
exceed	/ɪkˈsiːd/	*v.*	to be greater than a particular number or amount	超过
unprecedented	/ʌnˈpresɪdentɪd/	*adj.*	never having happened before, or never having happened so much	前所未有的；史无前例的
displace	/dɪsˈpleɪs/	*v.*	to force people to move away from their homes to another place	迫使离开家园
precipitation	/prɪˌsɪpɪˈteɪʃən/	*n.*	rain, snow, etc. that falls; the amount of this that falls	降水；降水量（包括雨、雪、冰等）
ice sheet				冰原

nitrogen oxide				氮氧化物
greenhouse gas emission				温室气体排放
hydroelectric power				水力发电
tropical cyclone				热带气旋
storm surge				风暴潮

Further Reading: Are Humans to Blame for Global Warming?

scrutinize	/ˈskruːtənaɪz/	*v.*	to examine someone or sth. very carefully	仔细查看
profligate	/ˈprɒflɪɡət/	*adj.*	using money, time, materials, etc. in a careless way	挥霍的
numerical	/njuːˈmerɪkəl/	*adj.*	relating to numbers; expressed in numbers	数字的
desiccated	/ˈdesɪkeɪtɪd/	*adj.*	completely dry	干涸的
rational	/ˈræʃənəl/	*adj.*	based on reason rather than emotions	合理的
gigaton	/ˈɡɪɡəˌtʌn/	*n.*	a unit of mass equal to one billion metric tons or one trillion kilograms	十亿吨
equilibrium	/ˌiːkwəˈlɪbriəm/	*n.*	a balance between different people, groups, or forces that compete with each other, so that none is stronger than the others and a situation is not likely to change suddenly	平衡
fossil fuel				化石燃料

water vapor 水蒸气

upper stratosphere 高平流层

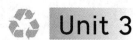

Carbon Neutrality

We must now agree on a binding review mechanism under international law, so that this century can credibly be called a century of decarbonization.

—*Angela Merkel*

Phase I First Sight

Video 1 Carbon Neutrality

New Words

offset /ˈɒfset/ v. 补偿；抵消

renewable /rɪˈnjuːəbəl/ adj. 可再生的

venue /ˈvenjuː/ n. 举办场所

renovate /ˈrenəveɪt/ v. 翻新；修复

streamline /ˈstriːmlaɪn/ v. 使效率更高

clear-cut /ˌklɪəˈkʌt/ adj. 明确的；明显的

loop /luːp/ n. 循环；环形

regenerative /rɪˈdʒenərətɪv/ adj. 再生的；更生的

1. Watch the video and choose the best answer.

(1) Which of the following is the first step to reduce the impacts on the environment?

    A. Carbon neutral.

    B. Net-zero emissions.

    C. Carbon negative.

    D. Carbon positive.

(2) To be carbon neutral, the amount of greenhouse gas released should be _____.

    A. reduced by credit companies

    B. matched by reductions elsewhere

    C. removed by carbon avoidance projects

    D. compensated for by emission companies

Unit 3　Carbon Neutrality

(3) An international target of net-zero emissions by 2050 set at the Paris Climate Summit _____.

　A. is agreed on by all countries

　B. is expected to remove the total amount of carbon dioxide in the atmosphere

　C. is expected to limit global warming to 1.5 °C this century

　D. is to achieve emission offsetting and carbon credits

(4) Which of the following statements is **NOT** true?

　A. Climate positive is at the top of our climb to minimal environmental impact.

　B. To be net-zero emissions, the indirect emissions are taken into account.

　C. Carbon neutral is not just about the balance between carbon-emitting and carbon-absorbing.

　D. Carbon negative is not just to avoid doing damage but to improve the natural world.

2. Watch the video again and complete the sentences with the words you hear.

(1) Buying credits helps fund carbon _____ projects around the world, like renewable energy generation or forest protection.

(2) Carbon neutral works for individuals too, and even _____ events.

(3) Carbon neutral doesn't take into account less _____, indirect emissions.

(4) A company (or a country, or an individual) works out its total emissions and puts these on one side of the _____.

(5) Less obvious indirect emissions are _____ in.

(6) The first step might be that the office and factory are _____ to high efficiency standards.

(7) Finally, the fleet of vehicles becomes _____.

(8) Climate positive action goes beyond net-zero and _____ the balance in favor of carbon removal.

3. Answer the following questions according to the video.

(1) What is the difference between carbon neutrality and net-zero emissions?

(2) What is the aim of carbon negative?

Video II "Explaining Shanghai" — Carbon Reduction

New Words

aggravate /ˈægrəveɪt/ v. 使恶化；使严重

consensus /kənˈsensəs/ n. 一致看法；共识

photovoltaic /ˌfəʊtəʊvɒlˈteɪɪk/ adj. 光电的；光伏发电的

sanitation /ˌsænɪˈteɪʃən/ n. 卫生设备

refuel /ˌriːˈfjuːəl/ v. 补充燃料；加燃料

integral /ˈɪntɪɡrəl/ adj. 必需的；不可或缺的

1. Watch the video and judge whether the following statements are TRUE or FALSE.

(1) _____ Slowing global warming by reducing greenhouse gases is agreed on among all countries in the world.

(2) _____ Carbon emissions per unit of Shanghai's GDP has dropped by 70 percent.

(3) _____ The percentage of coal in primary energy consumption in Shanghai will drop to 30 percent by 2025, down from the 2005 level.

(4) _____ China surpassed the EU as the world's largest carbon market.

(5) _____ The number of new-energy vehicles in Shanghai accounts for nearly half of all automobiles.

2. Watch the video again and fill in the blanks with the words you hear.

The climate-related extremes are not directly caused by global warming, but yet are (1) _____ by it. Reducing carbon dioxide and other greenhouse gas

Unit 3 Carbon Neutrality

emissions can slow down the global warming trend.

Shanghai expects its carbon emissions to (2) _____ by 2030. Work will be firstly done in the field of energy (3) _____. At the same time, Shanghai regards the carbon shift as an opportunity to (4) _____ traditional industries, and to take the lead in growing green industries. In 2021, China's carbon emissions trading market was (5) _____. Green transportation is the third key point. Last year the first (6) _____ energy bus line in China started operation in Shanghai. Energy-saving buildings are a(n) (7) _____ part of the city's low-carbon revolution. Shanghai will continue to be a GDP (8) _____ for China.

3. Answer the following questions according to the video.

(1) How will Shanghai achieve the goal of peaking carbon emissions by 2030?

(2) What has Shanghai done and will do in building green transportation?

Phase II Getting to Know

Warm-up Activity

Explain the following terms according to what you've explored before class.

♻ carbon neutrality   ♻ carbon credit   ♻ net-zero emissions

Active Reading

Carbon Neutrality and Net Zero

[1] Climate change is affecting the entire world, with extreme weather conditions such as drought, heat waves, heavy rain, floods, and landslides becoming more frequent. In order to **avert** the worst impacts of climate change and preserve a livable planet, global temperature increase needs to be limited to 1.5 °C above pre-industrial levels. Currently, the Earth is already about 1.1 °C warmer than it was in the late 1800s, and emissions continue to rise. To keep global warming to no more than 1.5 °C — as called for in the Paris Agreement — greenhouse gas emissions must peak before 2025 at the latest, decline 43 percent by 2030, and reach carbon neutrality by 2050.

What Is Carbon Neutrality?

[2] Carbon neutrality was the *New Oxford American Dictionary*'s word of the year in 2006 and since then, has been **catapulted** into the mainstream world. By definition, carbon neutrality is the balance between emitting carbon and absorbing carbon emissions through carbon sinks, any systems that absorb and store more carbon from the atmosphere than they emit, such as forests, soil, and oceans.

[3] There are several ways to achieve carbon neutrality and often a combination of methods is required. One common approach is to reduce emissions through the

use of **renewable energy** sources, such as solar and wind power. This can involve investing in renewable energy projects, installing renewable energy technology on a building or facility, or integrating energy storage systems.

[4] Another way to achieve carbon neutrality is through carbon **offsetting**. This involves funding projects that remove CO_2 from the atmosphere, such as reforestation or carbon capture and storage projects. These projects can offset the emissions produced by an individual or organization, effectively neutralizing their carbon footprint.

What Does It Mean to Become Carbon Neutral?

[5] Businesses often speak about becoming carbon neutral. This means they're taking steps to remove the equivalent amount of CO_2 to what's emitted through activities across their supply chains, by investing in carbon sinks that absorb CO_2.

[6] However, it's not just businesses that can strive for carbon neutrality — we can all make contributions as individuals. Making more sustainable lifestyle choices can help reduce our carbon footprint and limit our overall environmental impact. This could mean using public transport over a personal vehicle, limiting food waste, recycling packaging and old clothes, and monitoring the carbon intensity of your home's power usage.

Difference Between Carbon Neutrality and Net Zero

[7] Carbon-neutral and net-zero emissions are two similar terms. In both cases, companies are working to reduce and balance their carbon footprint. While carbon neutrality is more related to balancing out the total amount of carbon emissions, net-zero emissions refer to the zero emissions of all greenhouse gases, such as methane (CH_4), **nitrous oxide** (N_2O), and other **hydrofluorocarbons** (HFCs).

[8] The distinguishing factor between the two claims hinges on compensated emissions (also known as offsets). A business can claim carbon neutrality by measuring its emissions and then offsetting the balance through financed projects outside of its value chain, without actually reducing its own emissions. Net-zero

emissions, on the other hand, compel companies to more meaningfully reduce value chain emissions. Thus, the ultimate goal of net-zero emissions is to completely stop the increase in atmospheric greenhouse gases and stabilize the concentration of these gases at a safe level.

Benefits of Carbon Neutrality and Net Zero

[9] What is the difference in benefits of a carbon neutrality or net-zero approach? In carbon neutrality, the frequently used tools of carbon credits and carbon offsetting are helpful means of **channeling** funds from large greenhouse gas producers to the development of **sustainable** projects, giving time for companies as they **transition** towards zero emissions. One common example of emissions offsetting is through a **reforestation** project, where trees are planted to absorb carbon dioxide from the atmosphere. The carbon that is absorbed by the trees is considered a "carbon offset" since it compensates for the emissions that were produced elsewhere. According to a study on reforestation and climate change, by planting over 500 billion trees, it is estimated that we can capture 205 gigatons (Gt) of carbon, reducing atmospheric carbon by 25 percent. This would be a significant feat.

[10] In contrast to carbon neutrality, a net-zero approach gets much closer to the heart of the issue of sustainability. A net-zero approach revolves around organizations examining their own practices and finding ways to decrease their absolute emissions through reducing energy use, increasing energy efficiency, and transitioning to renewable energy sources. Some examples include using renewable energy sources like solar and wind, **electrifying** vehicle fleets to electric power, and finding better ways to utilize batteries through second-life applications. Overall, a net-zero approach is beneficial because it goes beyond carbon neutrality by requiring companies to take more aggressive action to reduce their carbon footprint and eliminate their contributions to climate change.

Challenges of Carbon Neutrality and Net Zero

[11] What are the main issues with the carbon neutrality and net-zero emissions approaches? A challenge posed by carbon neutrality is that a carbon-neutral

organization does not necessarily reduce its overall emissions. Carbon neutrality, for instance, can be achieved by emissions offsetting and purchasing carbon credits. A popular way of emissions offsetting is through reforestation projects. While reforestation projects would be a significant means of carbon offsetting, such projects alone would not be sufficient to address climate change, nor are they a substitute for reducing fossil fuel emissions. An important **caveat** with reforestation is that trees take many years to grow, mature, and sequester carbon. A study on reforestation and climate change describes, for instance, that reforesting an area the size of the United States and Canada could take one or even two thousand years. And even once trees are planted, the process of reaching maturity can take a century. Thus, a challenge with reforestation efforts is the lengthy amount of time required for trees to absorb significant atmospheric carbon.

[12] The general challenge, of course, is that with the concept of carbon neutrality, an organization can increase its emissions within its operations so long as it somehow compensates for them elsewhere so that the net amount calculated is zero. An important concern to be aware of, then, is that this system can **perpetuate** large-scale production of greenhouse gases, not addressing and reducing these emissions.

[13] A challenge of a net-zero approach is that it is not **standardized** and can have very different practices depending on the organization and its intentions. While some organizations are indeed genuinely working towards sustainability, others may continue to delay such efforts and hide behind ambiguous **buzzwords** like carbon neutral and net zero.

(1,097 words)

Paris Agreement《巴黎协定》

2015年12月12日在第21届联合国气候变化大会（巴黎气候大会）上通过，2016年11月4日起正式实施。《巴黎协定》是已经到期的《京都议定书》的后续。《巴黎协定》的长期目标是将全球平均气温较前工业化时期上升幅度控制在2 ℃以内，并努力将温度上升幅度限制在1.5 ℃以内。

1. Judge whether the following statements are TRUE or FALSE.

(1) _____ Carbon neutrality is expected to be achieved by 2050 as called for in the Paris Agreement in order to limit global warming within 1.5 ℃.

(2) _____ Carbon sinks refer to any systems that can remove sufficient carbon from the atmosphere.

(3) _____ Carbon neutrality of an organization involves the balance through financed projects outside its value chain, without actually reducing its emissions.

(4) _____ Reforestation is believed to be a good substitute for actually reducing fossil fuel emissions.

(5) _____ Net zero in the text refers to net-zero emissions which require the absolute reduction of greenhouse gases emissions.

2. Complete the following sentences with the words or phrase in the box below.

| catapulted | avert | sustainable | perpetuate |
| transitioned | caveat | buzzword | hinges on |

(1) Obviously, sustainability is a big important _____, especially with climate change and other issues.

(2) The company has _____ to new management in the past year.

(3) The aim of the association is to _____ the skills of traditional furniture design.

(4) At the end of the advertisement, the company gives a(n) _____ that the drug has some possible side effects.

(5) Aluminum is considered to be the most _____ material due to the little energy it takes to recycle.

(6) The fate of the project _____ the decision of the council.

(7) The film _____ her to fame overnight.

(8) The question for Western Europe's leaders is how best to _____ such a disaster.

3. Match the following technical terms with their Chinese equivalents.

I	II
(1) carbon sink	( ) 碳足迹
(2) carbon offsetting	( ) 碳减排
(3) carbon neutrality	( ) 净零排放
(4) carbon credit	( ) 碳中和
(5) carbon footprint	( ) 重新造林
(6) carbon reduction	( ) 碳补偿
(7) reforestation	( ) 碳汇
(8) net-zero emissions	( ) 碳积分/碳信用/碳权

4. Translate the following paragraphs into Chinese.

In contrast to carbon neutrality, a net-zero approach gets much closer to the heart of the issue of sustainability. A net-zero approach revolves around organizations examining their own practices and finding ways to decrease their absolute emissions through reducing energy use, increasing energy efficiency, and transitioning to renewable energy sources.

A challenge of a net-zero approach is that it is not standardized and can have very different practices depending on the organization and its intentions. While some organizations are indeed genuinely working towards sustainability, others may continue to delay such efforts and hide behind ambiguous buzzwords like carbon neutral and net zero.

Phase III Engaging Yourself

Warm-up Activity

Explain the following terms according to what you've explored before class.

- carbon peak
- carbon intensity
- reforestation
- energy intensity

Further Reading

Achieve Dual Carbon Goals in a Balanced Way

[1] Energy in different forms, such as heat and electricity, fuels the economy and helps humans meet their basic needs. Although China has rich coal reserves, they are still finite in terms of quantity, and **extracting** coal is full of uncertainties, for it requires hard labor and has huge ecological and social implications. And this does not agree with the energy and resource conservation policy which China has been emphasizing on the road to modernization which emphasizes the harmony between humankind and nature.

[2] True, fossil fuel-powered economic growth has paved the way for China to become the world's second-largest economy, but it has also brought large carbon emissions. That's why all local governments have energy-conservation and water-recycling departments to achieve self-sufficiency in energy.

[3] The first binding energy intensity target was set for the 11th Five-Year Plan (2006−2010) period. The 13th Five-Year Plan (2016−2020) adopted a dual control policy for the total energy consumption and energy intensity, capping the total energy consumption at 5 billion tons of standard coal equivalent annually. And energy consumption per unit of GDP (energy intensity) was expected to decline by 15 percent from 2015 to 2020.

[4] On the global front, China rose to leadership position in the climate governance system when some other countries **shirked** their international responsibilities. China announced at the 75th annual session of the United Nations General Assembly[1] in 2020 that it will peak its carbon emissions before 2030 and realize carbon neutrality before 2060.

[5] The 14th Five-Year Plan (2021–2025) for National Economic and Social Development and the Long-Range Objectives Through the Year 2035[2] started **synchronizing** the management of energy consumption and carbon emissions. It requires energy use and carbon intensity to decrease by 13.5 percent and 18 percent by 2025, respectively, compared with 2020. Overall, carbon intensity is expected to reduce by more than 65 percent by 2030 compared with 2005.

[6] China is committed to meeting the medium-term decarbonization targets set for 2025 and 2030. The stakes are too high to fail and decisions made today will have far-reaching implications for the future. In fact, the country has learned the lessons of what resorting to last-minute measures for meeting the dual control energy consumption targets means.

[7] From late September to October in 2021, many parts of China experienced **power outages**, **power rationing**, and severe disruptions in production, leading to market instability due to irregular supplies and rising prices of raw materials, which affected people's daily lives.

[8] And then the severe drought in the summer of 2022 lowered the generation capacity of hydropower, making coal the **bedrock** of energy security. Coal is critical for **grid** stability, as it accounts for about 70 percent of **peak load provision**.

---

1 the United Nations General Assembly：联合国大会，简称"联大"，成立于1945年10月24日，是由联合国全体193个会员国共同组成的主要审议、监督和审查机构。

2 The 14th Five-Year Plan (2021–2025) for National Economic and Social Development and the Long-Range Objectives Through the Year 2035：2021年3月11日，十三届全国人大四次会议表决通过了关于《中华人民共和国国民经济和社会发展第十四个五年规划和2035年远景目标纲要》的决议。

⑨ Since energy security, economic and social stability, and carbon reduction are all desirable goals, they need to be synchronized. Dual control of total energy consumption and intensity mainly focuses on the power sector and large end energy users. But decarbonization demands a **holistic** approach involving energy consumption, diversification of energy source, **structural upgrading of industries** to enable them to move up the value chains, reducing the economy's dependence on heavy and construction industries, electrification of transportation and industry, modernizing agriculture, and ecological and land preservation and restoration to increase carbon sinks.

⑩ Thus, dual control of total carbon emissions and energy intensity addresses the environment/climate-economy **dichotomy**. It has the potential to guide the formulation and implementation of a holistic and forward-looking long-term development strategy that is politically **viable**, technically functional, administratively operable, and financially feasible.

⑪ The dual energy policy provides a good basis for **aligning** economic and environmental/climate objectives and stimulating collective action. For example, power generators, grid companies, and local governments now have the **incentive** to work together to increase renewable energy generation and storage by overcoming challenges such as the **curtailment** of wind power and reluctance to share transmission **infrastructure**. It could also give rise to innovative measures to reform the power markets and differentiate customers by their preferences for energy mixes.

⑫ Upgrading of traditional energy and carbon intensive industries such as steel, **non-ferrous** metals, oil refining, chemicals, and building materials are already underway, and the efforts to reduce carbon emissions instead of direct energy consumption will provide more flexibility for and minimize **disruption** in the industrial and transportation systems.

⑬ China's new energy policy could help supply the missing links for the interconnected environmental, climate, economic and social issues, address the incompatibility in policy measures, and help realize the dual carbon target of peaking emissions before 2030 and achieving carbon neutrality before 2060 **via** balanced, if

Unit 3　Carbon Neutrality

not **optimal**, approaches.

⑭ China has invested heavily in research and development for finding technical solutions to energy-related problems. The key challenge lies in building an all-**encompassing** carbon emissions **inventory** for economic and human activities and institutions and policies enabling collective action based on monitoring, reporting, and verifying the volume of carbon emissions.

(801 words)

Five-Year Plan 五年规划（原称"五年计划"）

　　全称为《中华人民共和国国民经济和社会发展五年规划纲要》，是中国国民经济计划的重要部分，属长期计划，主要是对国家重大建设项目、生产力分布和国民经济重要比例关系等作出规划，为国民经济发展远景规定目标和方向。中国从1953年开始制定第一个"五年计划"。从"十一五"起，"五年计划"改为"五年规划"。

1. Read the text and choose the best answer.

(1) Which is **NOT** the reason why extracting coal is full of uncertainties?

　　A. Extracting coal requires hard labor.

　　B. Coal reserves are finite in terms of quantity.

　　C. Extracting coal has huge ecological and social implications.

　　D. Extracting coal does not agree with the energy and resource conservation policy.

(2) The dual control policy for the total energy consumption and energy intensity was adopted in the _____.

A. 11th Five-Year Plan

B. 12th Five-Year Plan

C. 13th Five-Year Plan

D. 14th Five-Year Plan

(3) Which of the following targets is set in the 14th Five-Year Plan?

A. Energy use is required to decrease by 18 percent by 2025.

B. Energy consumption per unit of GDP is expected to decline by 15 percent by 2025.

C. China will peak its carbon emissions before 2030 and realize carbon neutrality before 2060.

D. Carbon intensity is expected to reduce by more than 65 percent by 2030 compared with 2005.

(4) What does "China's new energy policy" in Paragraph 13 refer to?

A. Dual control of total energy consumption and intensity.

B. Dual control of total carbon emissions and energy intensity.

C. Dual target of peaking carbon emissions and achieving carbon neutrality.

D. Dual control policy for energy consumption and carbon intensity.

2. Replace the underlined parts with the words in the box below.

| extract | synchronizes | align | disruption |
| holistic | encompasses | viable | bedrock |

(1) The nuts are crushed in order to <u>obtain</u> oil from them.

(2) A flash lightning usually <u>happens with</u> a crash of thunder in summer.

(3) Whether such solar steam projects will be financially <u>feasible</u> for their developers is another matter.

(4) The job of an IT business analyst is to <u>coordinate</u> IT systems with changing business needs.

(5) A <u>global</u> perspective is taken on literary stylistics in addressing science fiction.

(6) It is a fruitful discussion that <u>includes</u> several different viewpoints.

(7) Some people believe that the family is the <u>basis</u> of society.

(8) A big jump in energy conservation could be achieved without much <u>disturbance</u> of anyone's standard of living.

3. Write a summary of the text in 120–150 words. You may use the given words or expressions in the box.

> the Five-Year Plan
> a dual control policy
> peak carbon emissions
> achieve carbon neutrality
> synchronizing/dual control of total carbon emissions and energy intensity
> environment-economy dichotomy   holistic development strategy
> market instability

Phase IV Commitment

1. Translate the following paragraph into English.

  中国在全球气候治理体系中占据了领导地位。2020 年第 75 届联合国大会上，中国宣布将在 2030 年之前达到碳排放峰值，2060 年之前实现碳中和。《中华人民共和国国民经济和社会发展第十四个五年规划和 2035 年远景目标纲要》开始同步管理能源消耗和碳排放，要求到 2025 年能源使用量和碳排放强度比 2020 年分别下降 13.5% 和 18%。总体而言，到 2030 年碳排放强度预计比 2005 年下降 65% 以上。

2. Work in pairs to discuss the following topics.

(1) Achieving carbon neutrality is an issue not just for businesses, but also for individuals. What contributions can college students make as individuals?

(2) There has been a long-standing argument about prioritizing the environment or the economy. Which do you think should receive the priority? Why?

3. Work in groups to make a presentation on any topic related to the theme of the unit.

Unit 3　Carbon Neutrality

Vocabulary

Active Reading: Carbon Neutrality and Net Zero

avert	/əˈvɜːt/	v.	to prevent sth. bad or dangerous from happening	防止；避免
catapult	/ˈkætəpʌlt/	v.	to push or throw sth. very hard so that it moves through the air very quickly	被猛掷，猛扔
offset	/ˈɒfset/	v.	to use one cost, payment, or situation in order to cancel or reduce the effect of another	补偿；抵消
hydrofluoro-carbon	/ˌhaɪdrəʊˈflwə-rəʊkɑːbən/	n.	a class of synthetic compounds that consist of hydrogen, fluorine, and carbon, and are widely used as refrigerants	一类含有氢、氟和碳的化合物，常见作制冷剂等；氟利昂
channel	/ˈtʃænəl/	v.	to control and direct sth. such as money or energy towards a particular purpose	导入；导向
sustainable	/səˈsteɪnəbəl/	adj.	able to continue without causing damage to the environment	可持续的；不破坏生态平衡的
transition	/trænˈzɪʃən/	v.	to change to a new state or start using sth. new	过渡；改变到新的状态
reforestation	/ˌriːfɒrɪˈsteɪʃən/	n.	the act of planting new trees in an area where there used to be a forest	重新造林
electrify	/ɪˈlektrɪfaɪ/	v.	to make sth. work by using electricity; to pass an electrical current through sth.	使电气化；使通电

caveat	/ˈkæviæt/	n.	a warning that sth. may not be completely true, effective, etc.	警告；告诫
perpetuate	/pəˈpetʃueɪt/	v.	to make a situation, an attitude, etc., especially a bad one, continue to exist for a long time	使永久化；使持续
standardize	/ˈstændədaɪz/	v.	to make all the things of one particular type the same as each other	使标准化；使符合标准
buzzword	/ˈbʌzwɜːd/	n.	a word or phrase, especially one connected with a particular subject, that has become fashionable and popular and is used a lot in newspapers, etc.	时髦术语；流行行话
renewable energy				可再生能源
nitrous oxide				一氧化二氮

Further Reading: Achieve Dual Carbon Goals in a Balanced Way

extract	/ɪkˈstrækt/	v.	to carefully remove a substance from sth. that contains it, using a machine, chemical process, etc.	提取；提炼
shirk	/ʃɜːk/	v.	to deliberately avoid doing sth. you should do, because you are lazy	推卸；逃避
synchronize	/ˈsɪŋkrənaɪz/	v.	to happen at exactly the same time, or to arrange for two or more actions to happen at exactly the same time	使同步；在时间上一致

bedrock	/ˈbedrɒk/	*n.*	strong base for sth. especially the facts or the principles on which it is based	牢固基础
grid	/grɪd/	*n.*	the network of electricity supply wires that connects power stations and provides electricity to buildings in an area	输电网；（输电线路）系统网络
provision	/prəˈvɪʒən/	*n.*	the act of supplying sb. with sth. that he or she needs or wants; sth. that is supplied	提供，供给，供应
holistic	/həʊˈlɪstɪk/	*adj.*	considering a person or thing as a whole, rather than as separate parts	整体的，全面的
dichotomy	/daɪˈkɒtəmi/	*n.*	the difference between two things or ideas that are completely opposite	一分为二；二分法
viable	/ˈvaɪəbəl/	*adj.*	that can be done; that will be successful	可实施的；切实可行的
align	/əˈlaɪn/	*v.*	to organize or change sth. so that it has the right relationship with sth. else	使一致
incentive	/ɪnˈsentɪv/	*n.*	sth. that encourages you to work harder, start a new activity etc.	激励；刺激
curtailment	/kɜːˈteɪlmənt/	*n.*	the act of limiting sth. or making it last for a shorter time	缩减；紧缩
infrastructure	/ˈɪnfrəstrʌktʃə(r)/	*n.*	the basic systems and structures that a country or an organization needs in order to work properly, for example, roads, railways, banks, etc.	（国家或机构的）基础设施；基础建设

non-ferrous	/nɒnˈferəs/	*adj.*	not containing or relating to iron	非铁或钢的
disruption	/dɪsˈrʌpʃən/	*n.*	a situation in which sth. is prevented from continuing in its usual way	扰乱；中断
via	/ˈvaɪə/	*prep.*	using a particular machine, system, or person to send or receive sth.	通过；凭借
optimal	/ˈɒptɪməl/	*adj.*	the best or most suitable	最佳的；最适的
encompass	/ɪnˈkʌmpəs/	*v.*	to include a wide range of ideas, subjects, etc.	包含；包括；涉及（大量事物）
inventory	/ˈɪnvəntri/	*n.*	a written list of all the objects, furniture, etc. in a particular building	（建筑物里的）清单；财产清单
power outage				供电中断
power rationing				电力配给
peak load				峰值负荷
structural upgrading of industries				产业结构升级

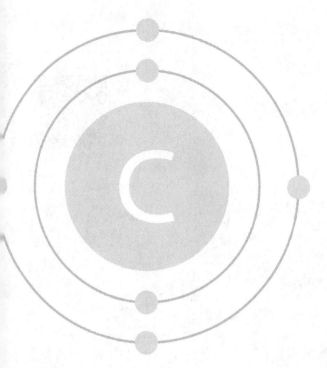

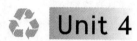

Carbon Trading

The Earth is a fine place and worth fighting for.

—*Anonymous*

Phase I First Sight

Video 1 What Is Carbon Trading?

New Words

allocation /ˌæləˈkeɪʃən/ *n.* 份额；分配

cap /kæp/ *v.* 限额；限制

intervention /ˌɪntəˈvenʃən/ *n.* 干预，介入；干涉

binding /ˈbaɪndɪŋ/ *adj.*（协议，承诺）有约束力的；必须遵守的

thermal /ˈθɜːməl/ *adj.* 热的；热量的

tackle /ˈtækəl/ *v.* 处理，应对（难题或局面）

1. Watch the video and choose the best answer.

(1) Which of the following statements is **NOT** true?

  A. Carbon trading is a market-based system that aims to reduce carbon emissions.

  B. Most of daily activities lead to greenhouse gas emissions.

  C. In voluntary offsets, consumers could pay a company to balance out their carbon footprint.

  D. All cap-and-trade systems have emissions limits determined by markets.

(2) If a company knows it has exceeded its allowance of emissions, it will need to _____.

  A. sell some carbon units on the carbon market

  B. ask for more carbon units from the government

  C. buy more carbon units from the carbon market

  D. make more carbon units itself

Unit 4　Carbon Trading

(3) The price of carbon is determined by _____.

　　A. non-profitable organizations

　　B. companies

　　C. consumers

　　D. supply and demand

(4) The European Union's Emission Trading System, the oldest active carbon market, was launched in _____.

　　A. 2003

　　B. 2005

　　C. 2007

　　D. 2009

(5) Carbon trading has the following features **EXCEPT** that _____.

　　A. it doesn't require much government intervention in the economy

　　B. it lets businesses figure out their own solutions

　　C. it has not received any critics in the past

　　D. it can efficiently drive decarbonization

2. Watch the video again and complete the sentences with the words you hear.

(1) Carbon trading is a legally binding scheme that caps total emissions and allows organizations to trade their _____.

(2) Carbon allowances, totaling up to the _____, are then allocated to companies and can be traded on a market.

(3) By assigning a price to damaging activities, the system provides a(n) _____ incentive for firms to reduce emissions.

(4) At the start of 2021, China _____ the world's largest carbon market for its thermal power industry.

(5) Cap-and-trade systems have been successful in _____ environmental problems in the past, including one covering sulphur dioxide emissions.

(6) Critics of carbon trading worry that countries facing economic difficulties might

be _____ to cheat.

(7) Another criticism of carbon markets is that developed countries, which have done most of the polluting to date, are able to _____ in low-carbon technology and have reoriented their economies to less carbon-intensive activities.

(8) The increasing popularity of cap-and-trade schemes, and the rising price of carbon allowances are _____ companies to consider their effect on the climate and has led to a reduction in emissions.

3. Answer the following questions according to the video.

(1) What is cap-and-trade scheme in carbon trading system?

(2) What happened in 2019 as governments across the world tightened environmental standards?

Video II How Do Carbon Markets Work?

New Words and Phrases

acid rain 酸雨

deterrent /dɪˈterənt/ *n.* 威慑因素；遏制力

lax /læks/ *adj.* 不严格的；不严厉的

power plant 发电厂

1. Watch the video and judge whether the following statements are TRUE or FALSE.

(1) _____ In 1991, the American government passed a law to force polluters to pay for their emissions by establishing a new market governed by a system called cap-and-trade.

(2) _____ In 1997, the international climate-change treaty, known as the Paris Agreement, suggested applying the concept of cap-and-trade to carbon.

Unit 4　Carbon Trading

(3) _____ A government sets a cap on the amount of CO_2 that can be emitted by an industry and splits the cap into permits that can be given or sold to firms.

(4) _____ A carbon market creates a race in which companies are motivated to buy more permits.

(5) _____ In theory in a cap-and-trade market, CO_2 emissions should fall, but in reality, they've continued to rise.

2. Watch the video again and fill in the blanks with the words you hear.

What's more, even if carbon is priced appropriately, the fines for (1) _____ permitted levels are sometimes (2) _____ low. In the EU, a fine can be as low as €100 per excess tonne. Considering that's not that much more than the price of a permit, it's hardly a(n) (3) _____.

Now in theory, regulators (4) _____ carbon prices and permits, but in practice, it can be very challenging. There are (5) _____ problems, direct emissions versus indirect emissions. There are questions of cheating; for example, you find in most markets around the world enforcement is (6) _____ and punishment is not great.

Since 2019, the EU has been taking steps to reduce the number of permits it (7) _____. Partly as a result, carbon prices in the EU are now (8) _____ record highs of over €60 per tonne.

3. Answer the following questions according to the video.

(1) What problem was America faced with towards the end of the 1980s?

(2) What would happen if carbon prices stay high enough?

Phase II Getting to Know

Warm-up Activity

Explain the following terms according to what you've explored before class.

- cap-and-trade
- regulated carbon market
- voluntary carbon market
- the Paris Agreement

Active Reading

Carbon Trading — What Is It and How Does It Work?

[1] The role of carbon markets is expected to increase rapidly in the coming years as nations continue to focus on sustainable economic development and drive towards a net-zero economy. Understanding how they work and the role they can play is critical.

From the 1990s to Today: How the Emission Reduction Strategy Began

[2] Carbon trading is a market-based system involving the trading of credits (of some form) aimed at reducing greenhouse gases that contribute to global warming, particularly carbon dioxide.

[3] The ultimate goal is to offset unavoidable carbon emissions by reducing emissions in countries or entities where it is most **cost-effective**.

[4] Carbon trading arose in the late 1990s with the signing of the Kyoto Protocol, which saw landmark commitments to cut greenhouse gases.

[5] An important element of the Kyoto Protocol was the establishment of the Clean Development Mechanism (CDM), a **flexible** market mechanism for carbon trading. Under the CDM industrialized economies can **implement** emission-reduction

Unit 4　Carbon Trading

projects in developing nations. These projects can **generate** saleable certified emission reduction (CER) credits. Industrialized nations can then use these CERs to meet part of their emission reduction targets under the Kyoto Protocol. As of June 2022, across the world, a total of 7,845 project activities were registered under the CDM.

[6] Several international treaties and conventions have refined and adjusted carbon trading since, including the Paris Agreement in 2015. Significantly, the Paris Agreement included a carbon trading "rulebook" in Article 6, but the details and mechanics were in negotiation.

[7] A critical breakthrough came in 2021 at the 26th UN Climate Change Conference in Glasgow (COP26). After years of stalemates, COP26 delivered crucial progress on the Article 6 "rulebook" including:

- setting new rules for regulating carbon accounting;
- creating a mechanism for countries to transfer carbon reductions;
- the establishment of a global carbon market **overseen** by a United Nations entity.

[8] The new UNFCCC Mechanism enables international emissions trading by both public and private actors.

[9] In November 2022, Egypt hosted the 27th UN Climate Change Conference (COP27), focused on moving from planning to implementation. This included the launch of the African Carbon Markets Initiative to help drive green development in the region.

Carbon Markets: Regulated and Voluntary

[10] Broadly, carbon markets fall into two categories: those introduced via regulatory **compliance** schemes and those that are voluntary.

[11] Participants of the regulated market trade emission allowances, in other words, they trade with permissions to pollute the atmosphere with greenhouse gases. The voluntary carbon market can be thought of as the **inverse** of this because it facilitates the trading of carbon credits afforded to individuals and corporations that have

77

removed, avoided or limited atmospheric greenhouse gases.

How Do Regulated Markets Work?

[12] Regulated or compliance Markets are created and regulated by **mandatory** national, regional, or international carbon reduction regimes, and are currently the main method used to obtain carbon reductions across **corporate** entities and industry **emitters**. Regulated markets can be split into two categories:

- Cap-and-trade schemes are the most common and work by setting a fixed limit on emissions through a set of permits released into the market via **auction** or distribution according to certain criteria.
- Baseline and credit mechanisms do not have a fixed limit on emissions, and companies trade offsets which are granted to companies that reduce their emissions more than they are otherwise obliged to. In this system, the items being traded are for past reduction in emissions as opposed to future pollution as seen in a cap-and-trade scheme.

[13] At a high level, a compliance market is used by companies and governments that are required by law to account for their greenhouse gas emissions. Regulatory carbon markets are supervised by state and international authorities.

[14] Features of cap-and-trade schemes include the following:

- Market participants are usually identified by governments based on sector, carbon emission intensity or size.
- Governments set an emissions limit (cap) on market participants and **issue** either by free allocation or auction a quantity of emission allowances consistent with that cap.
- An auctioning system is the preferred approach because it requires companies to purchase allowances and therefore further **incentivizes** companies to reduce their emissions.
- Over time this cap is reduced so that greenhouse gas emissions decline. To avoid heavy penalties at the end of each compliance period companies must surrender an allowance for every ton of greenhouse gas emitted.

- Companies that have reduced their emissions can either retain the spare allowances to cover future emissions or can sell their spare allowances to companies that will exceed their cap. By purchasing the spare allowances, the emitter is allowed to exceed their cap because they are effectively paying someone else to reduce their emissions on behalf of them.

[15] Examples of regulated markets include the following:

- The European Union's (EU) Emission Trading System (ETS) introduced in early 2005, the world's first. The EU ETS is a cap-and-trade scheme.
- China's national ETS introduced in 2021, the world's largest ETS.

How Do Voluntary Markets Work?

[16] Voluntary markets function outside of regulated markets and enable companies, government entities, and individuals to purchase carbon offsets on a voluntary basis with no intended or possible use for compliance purposes. Permits or credits are often purchased under voluntary schemes to achieve internal Corporate Social Responsibility (CSR) or public relations purposes or corporate initiatives. For example, an increasing number of companies are making commitments to reduce their own emissions, emissions associated with **supply chains**, and emissions produced using their products.

[17] Carbon credits, purchased voluntarily, enable companies to compensate for the emissions they have not been able to eliminate themselves and thus meaningfully contribute in the transition to global net-zero.

[18] More recently, the certification of carbon credits has come under a great deal of **scrutiny**. The success of these voluntary markets is **contingent** upon the accuracy and integrity of carbon credit certification.

[19] Features of voluntary markets include the following:

- They cover a wider variety of sectors than compliance markets which typically deal with high emitting industries such as energy production.
- An **overarching** aim of incentivizing private actors to finance projects that

- remove greenhouse gas emissions from the atmosphere.
- Regulation is often **outsourced** to non-governmental organizations (NGOs). These NGOs have created their own methodologies to certify emission reduction projects.
- Voluntary carbon markets are independent of regulatory markets and therefore companies under an emission cap cannot purchase voluntary carbon credits to meet their legal obligations.

[20] Examples of voluntary markets include the following:

- The aviation industry-wide reduction scheme: Under the Carbon Offsetting and Reduction Scheme for International Aviation (CORSIA) established by the International Civil Aviation Organization (ICAO), ICAO member states can voluntarily offset their aviation-based carbon emissions.
- Australia's carbon credit unit (ACCU) voluntary scheme.

(1,146 words)

Clean Development Mechanism (CDM) 清洁发展机制

《联合国气候变化框架公约》主导的发达国家与发展中国家之间的碳交易机制，意在支持减少温室气体在全球范围内的排放行动。

1. Judge whether the following statements are TRUE or FALSE.

(1) _____ Under the Clean Development Mechanism (CDM), industrialized economies can implement emission-reduction projects in developed nations.

(2) _____ After the 26th UN Climate Change Conference in Glasgow (COP26), a global carbon market supervised by a United Nations entity was established.

(3) _____ Participants of voluntary markets trade with permissions to pollute the

Unit 4　Carbon Trading

atmosphere with greenhouse gases.

(4) _____ The cap-and-trade scheme sets a flexible limit on emissions through a set of permits released into the market via auction or distribution.

(5) _____ Voluntary markets contribute to the transition to global net-zero as they enable companies to compensate for the emissions they have not been able to eliminate themselves.

2. Complete the following sentences with the words or phrase in the box below.

| overseeing | inverse | consistent with | implementation |
| mandatory | generated | offset | cost-effective |

(1) The savings on staff wages are _____ by the increased maintenance costs.

(2) Traditional publishers find it convenient and _____ to deliver documentation in electronic formats.

(3) Completing the project on time and under budget _____ a feeling of pride and accomplishment among the team.

(4) She will be responsible for _____ strategic and operational plans.

(5) Resources were made available to facilitate the _____ of these strategies.

(6) In short, there is a negative or _____ relationship between price and quantity demanded.

(7) The Council has made it _____ for all nurses to attend a refresher course every three years.

(8) The results are entirely _____ our earlier research.

3. Match the following technical terms with their Chinese equivalents.

I	II
(1) carbon permit allocation	(　　) 清洁发展机制
(2) Certified Emission Reduction	(　　) 碳排放权交易体系
(3) corporate social responsibility	(　　) 碳排放权分配
(4) clean development mechanism	(　　) 碳排放

81

(5) emission trading system （　　）自愿交易市场

(6) regulated market （　　）企业社会责任

(7) voluntary market （　　）供应链

(8) Kyoto Protocol （　　）核证减排量

(9) carbon emission （　　）配额交易市场

(10) supply chain （　　）京都议定书

4. Translate the following paragraphs into Chinese.

Carbon trading is a market-based system involving the trading of credits aimed at reducing greenhouse gases that contribute to global warming, particularly carbon dioxide.

The ultimate goal is to offset unavoidable carbon emissions by reducing emissions in countries or entities where it is most cost-effective. Broadly, carbon markets fall into two categories: those introduced via regulatory compliance schemes and those that are voluntary.

Regulated markets are created and regulated by mandatory national, regional, or international carbon reduction regimes, and are currently the main method used to obtain carbon reductions across corporate entities and industry emitters.

Voluntary markets enable companies to compensate for the emissions they have not been able to eliminate themselves and thus meaningfully contribute in the transition to global net-zero.

Unit 4　Carbon Trading

Phase III　Engaging Yourself

Warm-up Activity

Explain the following terms according to what you've explored before class.

- Clean Development Mechanism (CDM)
- certified emission reduction
- carbon emission allowance

Further Reading

China's National ETS — A Model

[1] A national carbon trading mechanism that has been put into operation in China could help promote the country's climate progress with market-based, cost-friendly solutions. The **endeavor** in the world's largest developing country could be a model for other nations as they explore viable climate actions.

[2] The Ministry of Ecology and Environment announced an **interim** regulation on the management of carbon trading in January 2021. Earlier, in December 2020, it published a document laying out the 2019—2020 allocation of carbon emission allowances for the power generation sector and a list of 2,225 companies that would be given emission allowances.

[3] These documents were **unveiled** as China is making increasingly intensified efforts to reduce carbon emissions. While addressing the general debate of the 75th session of the United Nations General Assembly via video in September 2020, President Xi Jinping announced that China aims to see CO_2 emissions peak before 2030 and achieve carbon neutrality before 2060.

[4] These documents **underscore** the fact that China's national carbon market has opened for business, according to the Environmental Defense Fund's China program.

83

[5] Carbon trading is the process of buying and selling permits to emit carbon dioxide or other greenhouse gases. If a company **curbs** its emissions significantly, it can sell **surplus** permits in the market. If it fails to limit its emissions, it has to buy unused allowances from other companies.

[6] The two documents published in 2020 indicate that the 2,225 power generation companies that saw their carbon emissions exceed their total free emission allocation limit between January 1, 2019 and the end of 2020 may have to buy permits in the market.

[7] The regulation unveiled in 2021 rules that companies could use China Certified Emission Reductions, which are generated within the China Greenhouse Gas Voluntary Emission Reduction Program, as offset credits. The program uses market-based incentives to reduce pollution. It basically revolutionized the ability of companies to drive down the emissions much faster than it would have taken place. But companies are only allowed to use such reductions to offset 5 percent of the permits they need to buy.

[8] The carbon trading program will be the first national environmental credit trading mechanism in China's environmental progression. It has replaced the European Union's carbon trading market, launched in 2005. Besides the EU, California in the United States and Quebec in Canada also **inaugurated** carbon trading markets in 2013 and combined them the following year.

[9] China started its pilot market in 2013. Considering the time taken for preparations, the Chinese market outpaces all the others. The country has accumulated the richest experiences in exploring how to establish a carbon trading market, especially in a developing country.

[10] In July 2021, China launched its national carbon trading market after years of preparations. Its **initial** phase includes over 2,000 power companies that collectively constitute approximately 6 percent of global emissions, or a little more than a fifth of China's entire emissions, which stand at an estimated 27 percent of the world's total. It will eventually scale to include other industries with time, thus increasing the

percentage of China's emissions that will be regulated through this structure. As of its launch, it's already the world's largest carbon trading market.

⑪ Ever since the national-level carbon trading system was rolled out, the basic structure of the carbon market has been shaped in China, helping companies reduce greenhouse gas emissions, accelerate their green transformation, and provide a **benchmark** for carbon pricing.

⑫ To enrich the financial characteristics of carbon trading, the China Securities Regulatory Commission released in April 2022 industrial standards for carbon-related financial products. More banks have implemented carbon emission rights pledge loans, and financial products linked to the carbon market have begun to emerge.

⑬ Systematic arrangements have been made, with management regulations for carbon trading and calculation methods for carbon emissions, verification, and settlement introduced. According to the overall plan of the Ministry of Ecology and Environment, from 2021 to 2025 during China's 14th Five-Year Plan period, the national carbon market will cover eight high-energy-consuming industries — power generation, iron and steel, construction materials, non-ferrous metals, petrochemicals, chemicals, paper manufacturing, and aviation, with a total of approximately 8,500 large emission-intensive enterprises. We will strive to transform our carbon emission management mechanism from one with a carbon intensity cap to one with an emission upper limit during the 15th Five-Year Plan period from 2026 to 2030.

⑭ This development is historic because it represents the most serious effort by any single country, to **combat** climate change. China is leading the way in this respect by setting the perfect model for others to follow. The experiences that China will obtain and subsequently share with others will be a priceless gift to humanity since everyone else can learn from its management of the world's largest carbon trading market.

(810 words)

> **China Certified Emission Reductions 中国核证自愿减排量**
>
> 将温室气体减排项目（可再生能源、林业碳汇、甲烷利用等）所吸收的二氧化碳，进行量化、核证、出售。2012年首次启动，允许在国家ETS之下进行交易的企业通过从某些减排项目（如造林和可再生能源发电）购买信用额度，以抵消至多5%的排放量。于2017年暂停。

1. Read the text and choose the best answer.

(1) In Paragraph 2, we learn that _____.

  A. an interim regulation on the management of carbon trading in China was announced in 2020

  B. the Ministry of Ecology and Environment published a document specifying the 2020–2021 allocation of carbon emission allowances

  C. more than 2,000 companies were listed to be given emission allowances in 2020

  D. the Ministry of Ecology and Environment published a document regarding carbon emission allowances after announcing regulation on the management of carbon trading

(2) According to Paragraph 7, the regulation announced in 2021 rules that _____.

  A. companies must buy permits in the market

  B. China Certified Emission Reductions could be used by companies as offset credits

  C. China Certified Emission Reductions work separately from the China Greenhouse Gas Voluntary Emission Reduction Program

  D. China Certified Emission Reductions could be used by companies as free emission allocation

(3) Which of the following statements is true?

  A. China aims to see CO_2 emissions peak before 2030 and achieve carbon

Unit 4 Carbon Trading

neutrality before 2050.

B. China started the pilot carbon trading market in 2005.

C. EU, California in the United States and Quebec in Canada launched their carbon trading markets in 2013.

D. China has accumulated rich experiences in exploring how to establish a carbon trading market, especially in a developing country.

(4) China's national-level carbon trading system has the following functions **EXCEPT** _____.

A. facilitating companies to emit greenhouse gas emissions

B. helping companies speed up their green transformation

C. excluding other industries in the carbon market

D. giving a benchmark for carbon pricing

(5) From 2021 to 2025 during China's 14th Five-Year Plan period, the national carbon market will _____.

A. issue industrial standards for carbon-related financial products

B. cover eight high-energy-consuming industries

C. complete the transformation of our carbon emission management mechanism to one with an emission upper limit

D. ban financial products linked to the carbon market

2. Replace the underlined parts with the words in the box below.

| inaugurated | endeavor | initial | underscore |
| curbing | combat | unveiled | benchmark |

(1) They suspected they would gain little from this particular <u>effort</u>, but anything that supplemented their income was worthwhile.

(2) The government has <u>revealed</u> plans for new energy legislation.

(3) A range of policies have been introduced aimed at <u>controlling</u> inflation.

(4) The moon landing <u>pioneered</u> a new era in space exploration.

(5) My <u>first</u> reaction was to decline the offer.

(6) Tests at the age of seven provide a <u>standard</u> against which the child's progress at school can be measured.

(7) The report <u>underlined</u> that the project enjoyed considerable support in both countries.

(8) To <u>fight against</u> inflation, the government raised interest rates.

3. **Write a summary of the text in 120–150 words. You may use the given words or expressions in the box.**

> carbon emissions   carbon trading
> national-level   offset
> pilot market   China Certified Emission Reductions
> initial phase   energy-consuming industries

Phase IV Commitment

1. Translate the following paragraph into English.

　　碳市场为处理好经济发展与碳减排关系提供了有效途径。全国碳排放权交易市场是利用市场机制控制和减少温室气体排放、推动绿色低碳发展的重大制度创新，也是落实中国二氧化碳排放达峰目标与碳中和愿景的重要政策工具。碳市场地方试点为全国碳市场建设摸索了制度，锻炼了人才，积累了经验，奠定了基础。

2. Work in pairs to discuss the following topics.

(1) Carbon markets are expected to contribute to a net-zero economy as nations continue to focus on sustainable economic development. Talk about the mechanism of the carbon trading system.

(2) China is leading the way to combat climate change. Talk about what future efforts we can make as a responsible nation.

3. Work in groups to make a presentation on any topic related to the theme of the unit.

Vocabulary

Active Reading: Carbon Trading — What Is It and How Does It Work?

cost-effective	/ˌkɑst ɪˈfektɪv/	*adj.*	giving the best possible profit or benefits in comparison with the money that is spent	有成本效益的
flexible	/ˈfleksəbəl/	*adj.*	able to change or suit new conditions or situations	灵活的
implement	/ˈɪmplɪment/	*v.*	to make sth. that has been officially decided start to happen or be used	实施；贯彻
generate	/ˈdʒenəreɪt/	*v.*	to produce or create sth.	产生
oversee	/ˌəʊvəˈsiː/	*v.*	to watch sb. or sth. and make sure that a job or an activity is done correctly	监督
compliance	/kəmˈplaɪəns/	*n.*	the practice of obeying rules or requests made by people in authority	遵从；服从；顺从
inverse	/ˌɪnˈvɜːs/	*n.*	the exact opposite of sth.	相反的事物
mandatory	/ˈmændətəri/	*adj.*	required by law	强制的
corporate	/ˈkɔːpərət/	*adj.*	connected with a corporation	公司的

Unit 4　Carbon Trading

emitter	/ɪˈmɪtə(r)/	*n.*	a person, an organization, or a country that produces and sends out a greenhouse gas such as carbon dioxide or a substance that pollutes the environment	发出者；发射者；排放者
auction	/ˈɔːkʃən/	*n.*	a public sale in which things are sold to the person who offers the most money for them	拍卖
issue	/ˈɪʃuː/	*v.*	to officially make a statement, give an order, warning, etc.	发行；发布；颁发
incentivize	/ɪnˈsentɪvaɪz/	*v.*	to encourage somebody to behave in a particular way by offering them a reward	激励（某人做某事）
scrutiny	/ˈskruːtəni/	*n.*	careful and thorough examination of sb. or sth.	仔细检查
contingent	/kənˈtɪndʒənt/	*adj.*	depending on sth. that may or may not happen	依情况而定的
overarching	/ˌəʊvərˈɑːtʃɪŋ/	*adj.*	very important, because sth. includes or influences many things	非常重要的，首要的
outsource	/ˈaʊtsɔːs/	*v.*	to arrange for somebody outside a company to do work or provide goods for that company	外包
supply chain				供应链

91

Further Reading: China's National ETS — A Model

endeavor	/ɪnˈdevə(r)/	*n.*	an attempt to do sth., especially sth. new or difficult	（尤指新的或艰苦的）努力；尝试
interim	/ˈɪntərɪm/	*adj.*	intended to be used or accepted for a short time only, until sth. or sb. final can be made or found	暂时的；过渡时期的
unveil	/ʌnˈveɪl/	*v.*	to show or introduce a new plan, product, etc. to the public for the first time	（使）公之于众，揭开；揭幕
underscore	/ˌʌndəˈskɔː(r)/	*v.*	to emphasize or show that sth. is important or true	强调
surplus	/ˈsɜːpləs/	*adj.*	more than is needed or used	过剩的；剩余的；多余的
curb	/kɜːb/	*v.*	to control or limit sth., especially sth. bad	控制
inaugurate	/ɪˈnɔːgjəreɪt/	*v.*	to introduce a new development or an important change	开创
initial	/ɪˈnɪʃəl/	*adj.*	happening at the beginning	最初的
benchmark	/ˈbentʃmɑːk/	*n.*	sth. that can be measured and used as a standard that other things can be compared with	基准
combat	/ˈkɒmbæt/	*v.*	to fight against an enemy	战斗

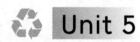

Unit 5

Green Materials

Bigger is not necessarily better, slower can be faster, and less can be more.

—*Anonymous*

Phase I First Sight

Video 1 The Miracle Material — Graphene

New Words and Phrases

graphene /ˈɡræfiːn/ *n.* 石墨烯

flaky /ˈfleɪkɪ/ *adj.* 易碎成小薄片的

disruptive /dɪsˈrʌptɪv/ *adj.* 创新性的；革新的；开拓性的

lead /led/ *n.* 铅；铅笔芯

brittle /ˈbrɪtl/ *adj.* 硬但易碎的；脆性的

hexagonal /hekˈsæɡənəl/ *adj.* 六边的；六角形的

pronounced /prəˈnaʊnst/ *adj.* 显著的；很明显的；表达明确的

field effect 场效应

pliable /ˈplaɪəbəl/ *adj.* 易弯曲的；柔韧的

go-to /ˈɡəʊtuː/ *adj.* （为解决某个问题或做某件事）必找的；首选的

semiconductor /ˌsemɪkənˈdʌkə(r)/ *n.* 半导体

bulletproof /ˈbʊlɪtpruːf/ *adj.* 防弹的

kevlar /ˈkevlɑː(r)/ *n.* 凯芙拉合成纤维（用于强化轮胎等橡胶制品）

electrochromic /ɪˌlektrəʊˈkrəʊmɪk/ *adj.* 电致变色的

concrete /ˈkɒŋkriːt/ *n.* 混凝土

dispersion /dɪˈspɜːʃən/ *n.* 分散剂；分散体

lesser /ˈlesə(r)/ *adj.* 次要的

Unit 5 Green Materials

1. Watch the video and choose the best answer.

(1) What's the commonality between diamond and graphite?

　　A. Both are one-atom material.

　　B. Both are forms of carbon.

　　C. Both are strong.

　　D. Both are two-dimensional material.

(2) Why do researchers believe that graphene could replace silicon?

　　A. Because graphene has a prominent field effect.

　　B. Because graphene has a noticeable electric field.

　　C. Because graphene has an obvious magnetic field.

　　D. All of the above.

(3) Which of the following descriptions about graphene is **NOT** true?

　　A. It is the thinnest material in the world.

　　B. It is extremely strong.

　　C. It is very flexible.

　　D. It is very brittle.

(4) Compared with conventional concrete, the new type of concrete obtained by adding graphene exhibits the quality of _____.

　　A. more powerful strength but lesser water resistance

　　B. better water resistance but lesser strength

　　C. more powerful strength and better water resistance

　　D. lesser strength and weaker water resistance

2. Watch the video again and complete the following sentences with the words you hear.

(1) Silicon is the _____ for semiconductor chips and the industry is based around this.

(2) Graphene has been used in _____ that had twice the stopping power of

the kevlar vest or 10 times better than a steel plate.

(3) _____ electrochromic glass can switch between solar heating and radiative cooling modes, with the largest tuning ranges in _____ radiation.

(4) Scientists developed a new method to add _____ graphene dispersion to a conventional concrete mix, and obtained a material with double the strength and four times the water resistance.

(5) The price of graphene varies greatly, ranging from _____ to _____ per pound according to the quality.

3. Answer the following questions according to the video.

(1) What is graphene? Describe it briefly.

(2) What kinds of applications of graphene are mentioned in the video?

Video II Materials Science for a Sustainable World

New Words and Phrases

microprocessor /ˌmaɪkrəʊˈprəʊsesə(r)/ *n.* 微处理器

Sweden /ˈswiːdən/ 瑞典

start from scratch 从头开始；白手起家

titanium /taɪˈteɪniəm/ *n.* 钛

Stockholm /ˈstɒkˌhəʊm/ 斯德哥尔摩（瑞典首都）

exhaust /ɪɡˈzɔːst/ *n.* 废气；尾气

porous /ˈpɔːrəs/ *adj.* 多孔的；透水的；透气的

sponge /spʌndʒ/ *n.* 海绵

pore /pɔː/ *n.* 毛孔；气孔；孔隙

optimize /ˈɒptɪmaɪz/ *v.* 使最优化；充分利用

Unit 5　Green Materials

> **insert** /ɪnˈsɜːt/ *v.* 插入；嵌入；添加
>
> **composite** /ˈkɒmpəzɪt/ *adj.* 合成的；混成的；复合的；*n.* 合成物；混合物；复合材料
>
> **prototype** /ˈprəʊtətaɪp/ *n.* 原型；雏形；最初形态

1. Watch the video and judge whether the following statements are TRUE or FALSE.

(1) _____ Making the world sustainable depends on the new materials that can be created or discovered.

(2) _____ Scientists usually begin from a similar end when they develop new materials.

(3) _____ Johanna has a clear idea about what the new materials will be used for when she begins creating them.

(4) _____ The super thin material created in Johanna's lab can help reduce greenhouse gas emissions.

(5) _____ This super thin material has already been used to cover an entire football field.

2. Watch the video again and fill in the blanks with the words you hear.

Different from Johanna's team, Martin's team's research process begins (1) _____ _____ with a clear goal: looking for the material that can store large amounts of hydrogen. Martin compares their research to cooking. You start off with a(n) (2) _____, which is just your idea, then you do experiments to try out the recipe. And then you analyze the results to see if we need to improve it and in which direction we should improve it.

To solve this problem, they are developing a(n) (3) _____ carbon material that will store hydrogen. The carbon material looks like a(n) (4) _____ with pores. Hydrogen atoms can be stored on the surfaces in the pores and the hydrogen takes up less space and cannot explode either.

97

The main goal of the research is to (5) _____ the material so that it can store as much hydrogen as possible and the hydrogen can be (6) _____ as efficiently as possible.

Martin's team is now trying to combine the carbon with metals in various (7) _____ materials that are able to release hydrogen already at much lower temperatures. Now researchers know where to move forward and further enhance these materials and get them closer to a(n) (8) _____.

3. Answer the following questions according to the video.

(1) What drawbacks does hydrogen have?
(2) What are researchers doing to solve the hydrogen storage problem?

Unit 5　Green Materials

Phase II　Getting to Know

Warm-up Activity

Explain the following terms according to what you've explored before class.

♻ metamaterials　　　♻ humanoid　　　♻ perovskite

Active Reading

Materials Science Breakthroughs

[1] To build today's smartphone in the 1980s, it would cost about $110 million, require nearly 200 kilowatts of energy (compared with 2 kw per year today), and the device would be 14 meters tall, according to Applied Materials[1] CTO Omkaram Nalamasu.

[2] That's the power of materials advances. Materials science has **democratized** smartphones, bringing the technology to the pockets of over 3.5 billion people. But far beyond devices and **circuitry**, materials science stands at the center of innumerable breakthroughs across energy, future cities, **transit**, and medicine. And at the forefront of COVID-19, materials scientists are forging ahead with **biomaterials**, **nanotechnology**, and other materials research to **accelerate** a solution.

[3] As the name suggests, materials science is the branch devoted to the discovery and development of new materials. It's an **outgrowth** of both physics and chemistry, using the periodic table as its grocery store and the laws of physics as its cookbook.

1　Applied Materials：应用材料公司，总部位于加利福尼亚州圣克拉拉市（Santa Clara）的一家美国公司，是仅次于荷兰 ASML 的全球第二大半导体设备供应商。

The Materials Genome Initiative[1]

[4] In June 2011 at Carnegie Mellon University[2], President Obama announced the Materials Genome Initiative, a nationwide effort to use open source methods and AI to double the pace of innovation in materials science. Obama felt this acceleration was critical to the U.S.' global competitiveness, and held the key to solving significant challenges in clean energy, national security, and human welfare. And it worked.

[5] By using AI to map hundreds of millions of different possible combinations of elements — hydrogen, **boron**, **lithium**, carbon, etc. — the initiative created an enormous database that allows scientists to play a kind of **improv** jazz with the periodic table.

[6] This new map of the physical world lets scientists combine elements faster than ever before and is helping them create all sorts of **novel** elements. And an **array** of new **fabrication** tools are further **amplifying** this process, allowing scientists and researchers to work at new scales and sizes, including the atomic scale, where materials are built one atom at a time.

[7] These tools have helped create the **metamaterials** used in carbon fiber composites for lighter-weight vehicles, advanced **alloys** for more durable jet engines, and biomaterials to replace human joints. Breakthroughs are also taking place in energy storage and **quantum computing**. In **robotics**, new materials are helping us create the artificial muscles needed for **humanoid**, soft robots — think *Westworld*[3] in your world.

[8] Let's **unpack** some of the leading materials science breakthroughs of the past decade.

---

1 Materials Genome Initiative：（美国）材料基因组计划

2 Carnegie Mellon University：（美国）卡内基梅隆大学

3 *Westworld*:《西部世界》（美国西部科幻电影）

Unit 5　Green Materials

Lithium-ion Batteries

⁹ The lithium-ion battery, which today powers everything from our smartphones to our autonomous cars, was first proposed in the 1970s. It couldn't make it to market until the 1990s, and didn't begin to reach maturity until the past few years.

¹⁰ An **exponential** technology, these batteries have been dropping in price for three decades, **plummeting** 90 percent between 1990 and 2010, and 80 percent since. **Concurrently**, they've seen an eleven-**fold** increase in capacity.

¹¹ But producing enough of them to meet demand has been an ongoing problem. Tesla has stepped up to the challenge: One of the company's Gigafactories[1] in Nevada[2] churns out 20 **gigawatts** of energy storage per year, marking the first time we've seen lithium-ion batteries produced at scale.

¹² Musk predicts 100 Gigafactories could store the energy needs of the entire globe. Other companies are moving quickly to integrate this technology as well: Renault[3] is building a home energy storage based on their Zoe[4] batteries, BMW's[5] 500 i3 battery packs are being integrated into the U.K.'s national energy grid, and Toyota[6], Nissan[7], and Audi[8] have all announced **pilot** projects.

¹³ Lithium-ion batteries will continue to play a major role in renewable energy storage, helping bring down solar and wind energy prices to compete with those of coal and gasoline.

1　Gigafactories:（特斯拉）超级工厂（世界最大电池工厂）

2　Nevada: 内华达州，位于美国西部。

3　Renault: 雷诺，法国汽车品牌，法国第二大汽车公司，最大的国有企业，世界十大汽车公司之一。

4　Zoe: 又称 Renault Zoe，雷诺·佐伊，法国雷诺汽车公司旗下的一款概念迷你电动车品牌。

5　BMW: 宝马，德国汽车品牌，总部设在德国慕尼黑。

6　Toyota: 丰田，日本最大的汽车公司，世界十大汽车工业公司之一。

7　Nissan: 日产，全称"日产汽车"，日本跨国汽车制造商，别名"尼桑"。

8　Audi: 奥迪，德国大众汽车集团旗下汽车品牌。

Graphene

[14] Derived from the same graphite found in everyday pencils, graphene is a sheet of carbon just one atom thick. It is nearly weightless, but 200 times stronger than steel. Conducting electricity and **dissipating** heat faster than any other known substance, this super-material has **transformative** applications.

[15] Graphene enables sensors, high-performance **transistors**, and even gel that helps neurons communicate in the **spinal cord**. Many flexible device screens, drug delivery systems, 3D printers, solar panels, and protective fabric use graphene.

[16] As manufacturing costs decrease, this material has the power to accelerate advancements of all kinds.

Perovskite

[17] Right now, the "conversion efficiency" of the average solar panel — a measure of how much captured sunlight can be turned into electricity — **hovers** around 16 percent, at a cost of roughly $3 per watt. Perovskite, a light-sensitive crystal and one of the new materials, has the potential to get that up to 66 percent, which would double what silicon panels can **muster**. What's more, perovskite's ingredients are widely available and inexpensive to combine. All these factors add up to affordable solar energy for everyone.

Materials of the Nano-World

[18] Nanotechnology is the outer edge of materials science, the point where matter manipulation gets nano-small — that's a million times smaller than an ant, 8,000 times smaller than a red blood cell, and 2.5 times smaller than a **strand** of DNA.

[19] Though invisible to the naked eye, nanoscale materials will integrate into our everyday lives, **seamlessly** improving medicine, energy, smartphones, and more. Progress has been surprisingly swift in the nano-world, with a great many of nano-products now on the market. Never want to fold clothes again? Nanoscale **additives** to **fabrics** help them resist wrinkling and staining. Don't do windows? Not a problem! **Nano-films** make windows self-cleaning, anti-reflective, and capable of

Unit 5　Green Materials

conducting electricity. Want to add solar to your house? We've got **nano-coatings** that capture the Sun's energy.

[20] Nanomaterials make lighter automobiles, airplanes, baseball bats, **helmets**, bicycles, luggage, power tools — the list goes on. Researchers at Harvard built a nanoscale 3D printer capable of producing miniature batteries less than one millimeter wide. And if you don't like those bulky VR **goggles**, researchers are now using nanotech to create smart contact lenses with a **resolution** six times greater than that of today's smartphones.

[21] On the environmental front, scientists can take carbon dioxide from the atmosphere and convert it into super-strong carbon nanofibers for use in manufacturing. If we can do this at scale — powered by solar — a system one-tenth the size of the Sahara Desert could reduce CO_2 in the atmosphere to pre-industrial levels in about a decade.

Concluding Thought

[22] With the help of artificial intelligence and quantum computing over the next decade, the discovery of new materials will accelerate exponentially. Advanced technologies start with re-designing materials — the **invisible enabler** and **catalyst**. The impact of the very, very small is about to get very, very large. Our future depends on the materials we create.

(1,089 words)

1. Judge whether the following statements are TRUE or FALSE.

(1) _____ Materials science plays a crucial role in countless advances in different fields.

(2) _____ The Materials Genome Initiative has made the U.S. a world leader in materials science.

(3) _____ The enormous database created by MGI enables scientists to create any new elements with the periodic table.

(4) _____ Perovskite may be quite promising in energy conversion and storage.

(5) _____ Without AI or quantum computing, discovery of new materials is impossible.

2. **Complete the following sentences with the words or phrases in the box below.**

| churn out | seamlessly | plummeting | amplify |
| democratize | at scale | dissipate | forging ahead |

(1) All further security and ticketing checks are performed _____ with facial recognition.

(2) Microsoft and Open AI are entering a new phase of their joint mission to _____ AI, i.e., to bring advanced artificial intelligence to the masses.

(3) It was recently announced that a new robot blacksmith is expected to _____ complex aircraft spare parts.

(4) Analysts predict that an oversupply of vehicles will lead to a price war that sends prices _____.

(5) The sandstorm that had ravaged most parts of China was expected to weaken and _____ by Tuesday evening.

(6) Robotics, automation, and AI — these innovations may promise to revolutionize healthcare _____.

(7) Businesses around the world keep _____ with expansions despite economic concerns and supply chain challenges.

(8) A latest study published in *Nature* found that high humidity levels can _____ heat risks in urban climates.

3. **Match the following technical terms with their Chinese equivalents.**

I	II
(1) biomaterial	(　) 纳米涂层
(2) quantum computing	(　) 生物材料
(3) humanoid	(　) 钙钛矿
(4) lithium-ion battery	(　) 仿真机器人
(5) resolution	(　) 石墨烯

(6) graphene   (　　) 锂离子电池

(7) high-performance transistor   (　　) 量子计算

(8) perovskite   (　　) 高性能晶体管

(9) nano-film   (　　) 纳米薄膜

(10) nano-coating   (　　) 分辨率

4. Translate the following paragraphs into Chinese.

Materials science has democratized smartphones, bringing the technology to the pockets of over 3.5 billion people. But far beyond devices and circuitry, materials science stands at the center of innumerable breakthroughs across energy, future cities, transit, and medicine. And at the forefront of COVID-19, materials scientists are forging ahead with biomaterials, nanotechnology, and other materials research to accelerate a solution.

These tools have helped create the metamaterials used in carbon fiber composites for lighter-weight vehicles, advanced alloys for more durable jet engines, and biomaterials to replace human joints. Breakthroughs are also taking place in energy storage and quantum computing. In robotics, new materials are helping us create the artificial muscles needed for humanoid, soft robots — think *Westworld* in your world.

Phase III Engaging Yourself

Warm-up Activity

Explain the following terms according to what you've explored before class.

- Materials Genome Engineering
- nanomaterial
- start-up
- data-mining

Further Reading

What's China Doing in Materials Science?

Hard Data

[1] 1980s: Chinese universities and research institutes develop 23 small-scale materials databases with financial support from the national government. They are used and updated infrequently.

[2] 2000: China **launches** two national, centralized materials databases involving 18 research institutes. For the first time, data are collected and entered in a standardized format.

[3] 2016: Policymakers invest in developing databases and big-data technology for China's Materials Genome Engineering[1] project, which **echoes** the Materials Genome Initiative launched by then-U.S. president Barack Obama in 2011.

[4] The Materials Genome Engineering (MGE) project is a national one-billion-yuan (U.S.$150-million) program launched in 2016 to revolutionize the speed and efficiency with which China can develop new materials. The program aims to **shepherd** advanced materials science into industrial applications. The name is intended

---

1 Materials Genome Engineering：材料基因工程

Unit 5　Green Materials

to draw comparisons to biological superprograms such as the Human Genome Project. In essence, China wants to make better use of the information stored in the country's databases about the behaviours of materials, so that new materials can be developed.

[5] One goal of the MGE is to develop a centralized, intelligent **data-mining** software platform that can offer instant feedback to companies involved in car manufacturing, steel-making, and shipbuilding on how materials behave. The final goals of the MGE is to produce better materials more quickly that cost less.

A Clear Funding Plan

[6] Funding for materials science in China has **quadrupled** since 2008, and the field receives the second highest level of funding from the National Natural Science Foundation of China[1] (NSFC), behind only medical sciences. The volume of China's materials-science research has grown correspondingly. According to data from the Web of Science[2], the number of papers on the topic more than **tripled** between 2006 and 2017, to around 40,000, and around one in every nine papers published by a Chinese researcher in 2015 was in materials science.

[7] Since 2006, China's scientific research and development (R&D) funding has been guided by a national plan to improve the country's level of innovation by 2020. The plan includes the realization of ambitious research and development projects, such as Moon exploration and the development of China's first domestically designed passenger aircraft. These goals are designed to **spur** technological breakthroughs and improve the country's economic prospects. Materials science is crucial to their success.

[8] In 2018, the NSFC pumped more than 2 billion yuan into 701 projects, including the MGE and work on nanotechnology and advanced electronic materials. In the same year, the Ministry of Science and Technology announced total funding

---

1　National Natural Science Foundation of China：国家自然科学基金委员会，简称"自然科学基金委"，成立于1986年，隶属于科学技术部，专门负责管理国家自然科学基金。

2　Web of Science：科学网，大型综合性、多学科、核心期刊引文索引数据库。

of more than 1.6 billion yuan for six special projects, which also covered nanotechnology.

⑨ China now publishes more high-impact research papers than any other country in 23 fields with clear technological applications, including batteries, semiconductors, new materials, and biotechnology. And in November, a **start-up** called Qingtao Energy Development, begun in 2014 by Ph.D. graduates from Tsinghua University, Beijing, announced that it had developed the country's first solid-state battery production line.

Talent Recruiting Initiative

⑩ In 2014, **geophysicist** Ho-Kwang Mao began splitting his time between the Carnegie Institution in Washington, D.C. and Shanghai. After more than 50 years working in the United States, he was hoping to help China solve one of the country's most **pressing** R&D problems: How to improve fundamental research in terms of quality, not just quantity?

⑪ Ho-Kwang Mao, born in Shanghai, is the scientist who studies how materials respond to extreme pressure and has had 65 papers published in *Nature*[1] and *Science*[2]. He told Chinese officials that they should give him the money to start a lab to produce "truly transformative research". He explained that there would probably be no immediate breakthroughs. His only guarantee was that he would attract scientists from all over the world who had the same potential for productivity that he himself has demonstrated. "Give me the money and I'll give you the scientists," he said in 2008. He got the money and in 2013, the Center for High Pressure Science and Technology Advanced Research (HPSTAR) was established, with branches in Shanghai and Beijing. Mao's labs are not funded through the usual central bodies, such as the Ministry of Science and Technology, but directly by the Ministry of Finance.

---

1 *Nature*：《自然》，世界上历史最悠久、最有名望的科学期刊之一。

2 *Science*：《科学》，世界顶级权威学术期刊之一。

Unit 5　Green Materials

⑫ Mao has attracted **staff scientists** from countries including Canada, Egypt, Germany, Japan, Russia, the United Kingdom, and the United States. He was allowed to try a new system that gives scientists total freedom to pursue transformative science in their own way with minimal supervision, reviews, and evaluations.

Fostering a Sustainable Graphene Industry

⑬ China recognized the contribution that nanoscience discoveries could make to its own scientific, technological, and economic development early on. In 2003, the Chinese Academy of Science and the Ministry of Education established the National Center for Nanoscience and Technology[1]. China now has the highest number of graphene businesses in the world — nearly 3,000, accounting for around two-thirds of global production.

⑭ But there's a problem: Almost all these producers are small or medium-sized companies that will lack funding in the long term unless they find a sustainable business model.

⑮ In 2013, an organization was formed to enable these companies to work together and grow. The good thing was there were lots of small companies working on graphene or graphene-related technologies. The bad thing was that quite a few of them couldn't keep going. In the same year, the China Innovation Alliance of the Graphene Industry[2] (CGIA) was founded, bringing together universities, institutes, and companies in an attempt to improve the situation.

⑯ The CGIA put together a database of graphene-related projects, categorized by their development stage: lab, pilot, and commercialization. Projects at each stage receive a different type of support. Lab projects are tracked and given an **incubator** when ready; at the pilot stage, the CGIA helps find investment; and at the commercialization stage, it invites project teams to present their work in front

---

1　National Center for Nanoscience and Technology：国家纳米科学中心

2　China Innovation Alliance of the Graphene Industry：中国石墨烯产业技术创新战略联盟

of potential investors and government representatives. The alliance has an annual operating budget of 8 million yuan.

[17] The next stage for the CGIA is to tackle one of the industry's biggest problems — the uncontrollable quality of graphene products. The effort in this area was the launching of the China International Graphene Industry Union[1] in 2016, which is now developing the country's first set of standards. The Union is also working on reducing technical barriers to trade and joint R&D with international teams such as the International Electrotechnical Commission[2] in São Paulo, Brazil.

[18] Right now Chinese scientists and researchers are working on setting up an international standardization evaluation committee to speed up the technology transfer from lab to industry. What China's materials science needs now is exposure to the rest of the world's scientists and work together as a global group.

(1,106 words)

1. Read the text and choose the best answer.

(1) Which of the following has received the most funding from NSFC since 2008?

　　A. Materials science.

　　B. Medical science.

　　C. Moon exploration.

　　D. Domestically designed passenger aircraft.

(2) Which of the following is **NOT** an effort at the national level?

　　A. Moon exploration.

　　B. Domestically designed passenger aircraft.

　　C. The establishment of HPSTAR.

　　D. The first solid-state battery production line.

1　China International Graphene Industry Union：中国国际石墨烯产业联盟

2　International Electrotechnical Commission：国际电工委员会，国际性的标准制定组织，负责制定电气、电子和相关技术领域的国际标准。

(3) The acceleration of technology transfer from lab to industry will be chiefly dealt with by _____.

   A. an international standardization evaluation committee

   B. China International Graphene Industry Union

   C. China Innovation Alliance of the Graphene Industry

   D. International Electrotechnical Commission

(4) To ensure a sustainable development for China's graphene industry, CGIA made the following efforts **EXCEPT** _____.

   A. offering different types of help to graphene-related projects based on their development stage

   B. establishing a specialized organization to develop the country's first set of standards of graphene industry

   C. reducing technical barriers to joint R&D with international teams

   D. investing 8 million yuan in pilot stage projects every year

2. **Replace the underlined parts with the words or phrases in the box below.**

   | start-ups | draw comparisons to | spur | echoed |
   | launched | in essence | shepherded | pressing | transformative |

(1) The rivalry between the world's major powers in the future is, essentially, the competition for scientific and technological strengths, particularly the competition for talent.

(2) With a 1.4 billion population, a massive user market, fierce rivalry among entrepreneurs, and enormous subsidies, China has been rated the most fertile ground for companies in the initial stages of business.

(3) The squad (小分队) became disoriented soon after entering the jungle. Fortunately, they had a local guide, who eventually guided the company out of the maze-like jungle.

(4) The house must undergo revolutionary renovations if it is to remain habitable.

(5) Borophene bears similarities to graphene in many aspects and is even much

better, particularly in hydrogen capture and retention properties.

(6) The thriving of tourism has <u>boosted</u> equivalent growth in the hotel and leisure-related sectors.

(7) The final experimental data <u>supported</u> the previous inferences and assumptions.

(8) The Belt and Road Initiative, proposed in 2013 and officially <u>implemented</u> in 2015, is a platform to strengthen the exchanges and cooperation between China and other countries.

3. Write a summary of the text in around 120–150 words. You may use the given words or expressions in the box.

> materials science
> funding industrial applications
> fundamental research
> small and medium-sized
> technological transfer
> standards
> significant increase
> technological breakthroughs
> sustainable development
> development stages
> technical barriers

Phase IV Commitment

1. Translate the following paragraph into English.

  石墨烯从发现至今虽然不到二十年，但是它所带来的轰动却是其他材料无法比拟的。中国虽然不是石墨烯的发现者，但是其丰富的原材料、不俗的科研实力、充满活力的经济环境和利好的政策使中国石墨烯产业走在了世界前列。打通石墨烯的产业化路径、保持中国在全球石墨烯产业中的领先优势是中国政府、企业以及科研机构共同奋斗的目标。

2. Work in pairs to discuss the following topics.

(1) List a few frequently used new materials and talk about their applications in daily life.

(2) How do you understand the statement that materials are the symbol of the progress in human civilization?

3. Work in groups to make a presentation on any topic related to the theme of the unit.

Vocabulary

Active Reading: Materials Science Breakthroughs

CTO			chief technology officer	首席技术官
democratize	/dɪ'mɒkrətaɪz/	v.	to make sth. available to all people; to make it possible for all people to understand sth.	普及；使大众化
circuitry	/'sɜːkɪtri/	n.	a system of electric circuits	电路系统；电路；电路装置
transit	/'trænsɪt/	n.	the process of or a system for moving goods or people from one place to another	运送；运输；运载
biomaterial	/ˌbaɪəʊmə'tɪəriəl/	n.	a substance that can be used in someone's body to help with a disease, injury, or other medical condition	（适用于人身体内治疗疾病、损伤或其他医学状况的）生物材料
nanotechnology	/ˌnænəʊtek'nɒlədʒi/	n.	a science which involves developing and making extremely small but very powerful machines	纳米技术
accelerate	/ək'seləreɪt/	v.	to happen or make sth. happen sooner or faster	使增速；使加快；促进；提前
outgrowth	/'aʊtgrəʊθ/	n.	sth. that develops from sth. else as a natural result of it	自然发展（或结果）
boron	/'bɔːrɒn/	n.	a chemical element; a solid substance used in making steel alloys and parts for nuclear reactors	（化学元素）硼

Unit 5 Green Materials

lithium	/ˈlɪθiəm/	n.	a soft silver-white chemical element that is the lightest known metal used in batteries, and often combined with other metals	（化学元素）锂
improv	/ˈɪmprɒv/	n.	acting, singing, performing, etc. without preparing in advance what you will say first	即兴表演；即兴乐；即兴剧
novel	/ˈnɒvəl/	adj.	new and original, not like anything known or seen before	新颖的；新奇的
array	/əˈreɪ/	n.	a large group of things or people, especially one that is attractive or causes admiration or has been positioned in a particular way	（指非常有吸引力、令人赞赏并常以特定的方式排列的）一系列；一批；大量
fabrication	/ˌfæbrɪˈkeɪʃən/	n.	the act of producing a product, especially in an industrial process	制造；装配
amplify	/ˈæmplɪfaɪ/	v.	to increase the size or effect of sth.	增强；放大；扩大
metamaterial	/ˌmetəməˈtɪəriəl/	n.	a usually artificial material that exhibits special properties not normally found in nature	超材料
alloy	/ˈælɔɪ/	n.	a metal that consists of two or more metals mixed together	合金
robotics	/rəʊˈbɒtɪks/	n.	the study of how robots are made and used	机器人学；机器人技术
humanoid	/ˈhjuːmənɔɪd/	n.	a machine or creature that looks and behaves like a human	仿真机器人；类人生物

unpack	/ˌʌnˈpæk/	v.	to separate sth. into parts so that it is easier to understand	分析；剖析
exponential	/ˌekspəˈnenʃəl/	adj.	(of a rate of increase) becoming faster and faster	（增长率）越来越快的；呈几何数（增长）的
plummet	/ˈplʌmɪt/	v.	to suddenly and quickly decrease in value or amount	暴跌；速降
concurrently	/kənˈkʌrəntli/	adv.	at the same time	并发地；同时发生地；并存地
-fold	/fəʊld/		(in adjectives and adverbs) multiplied by; having the number of parts mentioned	乘以；……倍；由……部分组成
gigawatt	/ˈɡɪɡəˌwɒt/	n.	a unit of power equal to one billion watts	千兆瓦；十亿瓦特
pilot	/ˈpaɪlət/	adj.	done on a small scale in order to see if sth. is successful enough to do on a large scale	试验性的；试点的
dissipate	/ˈdɪsɪpeɪt/	v.	to gradually become less or weaker before disappearing completely, or to make sth. do this	（使）消散；消失；驱散
transformative	/trænsˈfɔːmətɪv/	adj.	causing a major change to sth. or someone, especially in a way that makes it or them better	彻底改观的；使大变样的；有改造能力的
transistor	/trænˈzɪstə(r)/	n.	a small electrical device containing a semiconductor, used in televisions, radios, etc. to control or increase an electric current	晶体管

Unit 5　Green Materials

perovskite	/pəˈrɒvskaɪt/	n.	a mineral that contains various types of metals that are used in industry	钙钛矿
hover	/ˈhɒvə(r)/	v.	to stay at or near a particular level	（在某一水平附近）徘徊
muster	/ˈmʌstə(r)/	v.	(also muster up sth.) to find as much support, courage, etc. as you can	聚集；激起（支持、勇气等）
strand	/strænd/	n.	a thin thread of sth., often one of a few, twisted around each other to make a string or rope	（线、绳、金属线、毛发等的）股；缕
seamlessly	/ˈsiːmləsli/	adv.	without any sudden changes, interruptions, or problems	无缝地；不停顿地；顺利地；连续地
additive	/ˈædətɪv/	n.	a substance added to another to strengthen or otherwise alter it for the purpose of improving the performance of the finished product	添加剂；添加物
fabric	/ˈfæbrɪk/	n.	cloth used for making clothes, curtains, etc.	织物；布料
nano-film	/nænəʊˈfɪlm/	n.	an extremely thin layer of sth. on a surface that is only one billionth of the stated unit	纳米薄膜
nano-coating	/nænəʊˈkəʊtɪŋ/	n.	an extremely thin layer of a substance covering a surface	纳米涂层

helmet	/ˈhelmɪt/	*n.*	a strong hard hat that soldiers, motorcycle riders, the police, etc. wear to protect their heads	头盔；防护帽
goggle	/ˈgɒgəl/	*n.*	a pair of tight-fitting eyeglasses with side shields worn to protect the eyes from hazards such as wind, glare, water, or flying debris	（用以挡光、防尘、防水等的）护目镜
resolution	/ˌrezəˈluːʃn/	*n.*	the ability of television, camera or computer screen, a microscope, etc. to show things clearly	（电视、相机、计算机显示屏、显微镜等的）清晰度；分辨率
invisible	/ɪnˈvɪzəbəl/	*adj.*	that cannot be seen	无形的；不可见的
enabler	/ɪˈneɪblə(r)/	*n.*	sth. or sb. that makes it possible for a particular thing to happen or be done	（使某事可能发生或完成的）促成者；赋能者
catalyst	/ˈkætəlɪst/	*n.*	a substance that makes a chemical reaction happen more quickly without being changed itself	（化学）催化剂
quantum computing				量子计算
spinal cord				脊髓

Further Reading: What's China Doing in Materials Science?

launch	/lɔːntʃ/	*v.*	to start an activity, especially an organized one	开始；从事；发起

Unit 5 Green Materials

echo	/ˈekəʊ/	*v.*	to repeat an idea or opinion because you agree with it	重复；附和
shepherd	/ˈʃepəd/	*v.*	to lead or guide a group of people somewhere, making sure that they go where you want them to go	带领；引导
data-mining	/ˈdeɪtəˌmaɪnɪŋ/	*n.*	the process of using a computer to examine large amounts of information, for example, about customers, in order to discover things that are not easily seen or noticed	数据挖掘（即指用计算机从资料中发掘资讯或知识）
quadruple	/ˈkwɒdrʊpəl/	*v.*	to increase and become four times as big or as high, or to make sth. increase in this way	（使）成四倍；（使）以四乘
triple	/ˈtrɪpəl/	*v.*	to increase by three times as much, or to make sth. do this	（使）成三倍；（使）增加两倍
spur	/spɜː(r)/	*v.*	to encourage a horse to go faster, especially by pushing it with special points on the heels of your boots	激励；鞭策；促进；加速
start-up	/ˈstɑːtʌp/	*n.*	a small business that has just been started, especially one whose work involves computers or the Internet	刚起步的新兴小型企业（尤指涉及电脑与互联网）
geophysicist	/ˌdʒiːəʊˈfɪzɪsɪst/	*n.*	a scientist who studies the physics of the Earth, including its atmosphere, climate, and magnetism	地球物理学家

119

pressing	/ˈpresɪŋ/	*adj.*	needing to be discussed or dealt with very soon	紧急的；急迫的
incubator	/ˈɪŋkjəbeɪtə(r)/	*n.*	an organization or place that aids the development of new business ventures, especially by providing low-cost commercial space, management assistance, or shared services	（新兴小企业的）孵化基地
staff scientist			（科研项目的）专职科研人员 / 科学家	

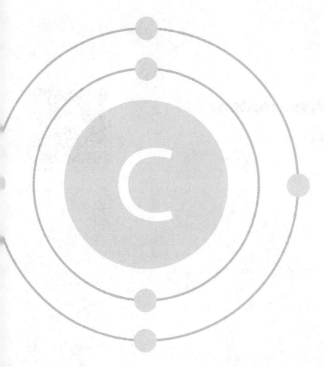

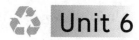 Unit 6

Clean Energy

A transition to clean energy is about making an investment in our future.

—*Gloria Reuben*

Phase I First Sight

Video 1 Advantages and Disadvantages of Renewable Energy

New Words

replenish /rɪˈplenɪʃ/ *v.* 补充；重新装满

geothermal /ˌdʒiːəʊˈθɜːməl/ *adj.* 地热的；地温的

maintenance /ˈmeɪntənəns/ *n.* 维护；保养

capacity /kəˈpæsəti/ *n.* 能力

disrupt /dɪsˈrʌpt/ *v.* 扰乱；使中断；妨碍

outlay /ˈaʊtleɪ/ *n.* 开支；费用

outweigh /ˌaʊtˈweɪ/ *v.* 重于；大于；超过

1. **Watch the video and choose the best answer.**

(1) How many advantages of renewable energy are mentioned in the video?

　　A. One.

　　B. Two.

　　C. Three.

　　D. Four.

(2) Which of the following statements about renewable energy is true?

　　A. Renewable energy projects can only bring clean environment to local areas.

　　B. Most renewable energy projects are located in the big cities.

　　C. Renewable energy needs no refuel.

Unit 6 Clean Energy

    D. Renewable energy facilities require equal maintenance as traditional power engines.

(3) Insufficient supply of electricity generated by renewable energies can be solved by _____.

    A. balancing different power sources

    B. reducing energy consumption

    C. building more energy facilities

    D. all of the above

(4) Which of the following does **NOT** explain why renewable energy can be unreliable?

    A. The weather can be unpredictable.

    B. It requires a massive financial expenditure.

    C. There is no sunlight at night.

    D. There may be droughts sometimes.

2. Watch the video again and complete the following sentences with the words you hear.

(1) Renewable energy comes from natural sources or processes that are constantly _____.

(2) Renewable energy facilities generally require less _____ than traditional generators and there is no need to _____.

(3) Renewable energies produce _____ or no waste products and don't _____ global warming. This means an overall positive impact on the natural environment.

(4) Unpredictable weather events make certain resources _____ and _____ technologies, affecting capacity to make energy from these resources.

(5) Renewable energy requires a huge financial _____ because the current cost of renewable energy technology is far _____ traditional fossil fuel generation.

(6) There are both _____ of using renewable energy. Do you think the advantages _____ the disadvantages?

3. Answer the following questions according to the video.

(1) What are the advantages of renewable energy?

(2) What are the disadvantages of renewable energy?

Video II Can 100 Percent Renewable Energy Power the World?

New Words and Phrases

flip /flɪp/ v. 快速翻转

ponder /ˈpɒndə(r)/ v. 沉思；考虑；琢磨

blast furnace （炼铁的）高炉；鼓风炉

radiate /ˈreɪdieɪt/ v. （使热、光、能量）辐射；放射；发散

quadrillion watt 千万亿瓦

hurdle /ˈhɜːdəl/ n. 难关；障碍

biomass /ˈbaɪəʊmæs/ n. 生物质能（燃料）

crisscross /ˈkrɪs krɒs/ v. （在……内）纵横交错；交叉；贯穿

astronomical /ˌæstrəˈnɒmɪkəl/ adj. 极其巨大的

resistance /rɪˈzɪstəns/ n. 阻力

megajoule /ˈmegəˈdʒuːl/ n. 兆焦

vessel /ˈvesəl/ n. （大型）船；舰

leap /liːp/ n. 巨变；剧增；跃进

1. Watch the video and judge whether the following statements are TRUE or FALSE.

(1) _____ At present, renewable energies only provide roughly 30 percent of our needs.

(2) _____ Global energy use is a diverse and complex system that requires an all-in-one solution.

(3) _____ Electricity is chiefly used for fixed devices in homes, businesses, and manufacturing, while liquid fuels for various forms of transportation.

(4) _____ Energy efficiency and transportation are two of the obstacles that restrict the large-scale use of solar energy.

(5) _____ Other forms of renewables are also faced with the similar problem of availability and location as solar energy.

2. Watch the video again and fill in the blanks.

In principle, a connected electrical energy network with power lines (1) _____ the globe would enable us to transport power from where it's generated to where it's needed. But building a system on this scale faces a(n) (2) _____ price tag.

Present-day power lines lose about 6–8 percent of the energy they carry because wire material (3) _____ energy through (4) _____. Longer power lines would mean more energy loss. Superconductors that can transport electricity without dissipation could be one solution. Unfortunately, such materials only work in low temperatures, which requires energy and (5) _____ the purpose. We'd need to discover new superconducting materials that operate at room temperature.

What about the all-important, oil-derived (6) _____? The scientific challenge is to store renewable energy in an easily transportable form. Even the best of lithium ion batteries can only store about 2.5 megajoules per kilogram, about 20 times less than the energy in one kilogram of gasoline.

To be truly competitive, car batteries would have to store much more energy without adding cost. The challenges only increase for bigger (7) _____, like ships and planes. To power a cross-Atlantic flight for a jet, we'd need a battery weighing about 1,000 tons. This, too, demands a technological (8) _____ towards new materials with higher energy density and better storage.

3. Answer the following questions according to the video.

(1) Human technology is advanced enough to capture all energy from renewables. List some factors that prevent solar energy from large-scale applications worldwide.

(2) In regard to clean or renewable energy, what are the challenges that automobile fueling faces?

Unit 6 Clean Energy

Phase II Getting to Know

Warm-up Activity

Explain the following terms according to what you've explored before class.

- clean energy
- green energy
- renewable energy
- biomass

Active Reading

All We Need to Know About Clean Energy

[1] Clean energy is energy that comes from renewable, zero emission sources that do not pollute the atmosphere when used, as well as energy saved by energy efficiency measures.

[2] There is a degree of **crossover** between clean energy and green or renewable energy sources, but they are not exactly the same. In order to understand the difference, it is worth understanding what it actually means.

What Does Clean Energy Mean?

[3] Clean energy is energy gained from sources that do not release air pollutants, while green energy is energy derived from natural sources. There is a subtle difference between these two energy types even though they are often spoken of as being the same. Renewable energy, however, is power generated from sources that are constantly replenished. Unlike fossil fuels and gas, these renewable energy resources, such as wind and solar energy, won't run out.

[4] However, while most green energy sources are renewable, not all renewable energy sources are seen as being green. For example, **hydropower** is a renewable

127

resource, but some would argue that it is not green, since the deforestation and industrialization related to the building of hydro dams can damage the environment. Similarly, while most clean energy sources are renewable, not all renewable energy sources are clean. For example, burning wood from sustainably managed forests can be renewable, but it is not clean since this releases carbon dioxide into the atmosphere.

[5] To be truly clean, the carbon cost of production and storage needs to be zero. The perfect clean energy mix occurs where green energy meets renewable energy, such as solar energy and wind energy.

[6] An easy way to remember the differences between these different energy types is:

Clean energy = clean air

Green energy = natural sources

Renewable energy = recyclable sources

Why Is It Important?

[7] Clean energy provides a variety of environmental and economic benefits. It can reduce the risk of environmental disasters, such as fuel **spills** or the problems associated with natural gas leaks. A diverse clean energy supply also reduces the dependence on imported fuels, and the associated financial and environmental costs this **incurs**.

[8] Renewable clean energy also has inherent cost savings, as there is no need to extract and transport fuels, such as oil or coal, as the resources replenish themselves naturally.

[9] Other industrial benefits of a clean energy mix is the creation of jobs to improve the infrastructure, manufacture clean energy solutions and install and maintain them. Renewable and clean energies are growth sectors as the world begins to move away from fossil fuels, meaning that more opportunities will arise in areas ranging from e-mobility to power generation and storage.

Unit 6　Clean Energy

⑩ Of course, the financial implications of clean energy are just part of the story, since the real incentive behind clean energy is creating a better future for the planet.

Major Types of Clean Energy

⑪ Clean energy can be obtained from a variety of sources which, when put together, could create solutions for all of our energy needs, from electricity generation to heating water and more, depending on the source of the energy.

⑫ Sunlight is the most abundant and freely available energy resource on the planet. The amount of solar energy that reaches the Earth in one hour is enough to meet the total energy requirements for the planet for an entire year. Solar energy can be used for heating and lighting buildings, generating electricity, and heating water directly. Solar panels are frequently used for small electric tasks, such as charging batteries. While many people use solar energy for small garden lanterns, this same clean energy technology can be scaled up to larger panels that are used to provide power for homes or other buildings. Even multiple solar panels are arrayed and installed to power entire towns.

⑬ Wind power is another plentiful source of clean energy. Wind power works by attaching a **windmill** to a generator which turns the turning of the windmill blades into power. This form of energy has been used for centuries to **grind** grain, pump water or perform other mechanical tasks, but is now being used more often to produce electricity. Onshore and offshore **windfarms** are becoming increasingly prevalent. But wind power can also be used on a much smaller scale to produce electricity, even to provide power for recharging mobile telephones.

⑭ Water is another clean resource with some surprising applications. Most obvious are hydroelectric power plants, which take the flow of water from rivers, streams or lakes to create electricity. Hydro or water power is one of the most commercially developed sources of clean energy. It is more reliable than either wind or solar power and allows for the easy storage of the energy that is generated. A less obvious use of water comes through **municipal** pipes in towns and cities. With lots of water running through pipes in homes each day, there is a move towards

129

harnessing this energy to help meet domestic and other power needs. As generators become smaller and less expensive to build, this use of municipal water is becoming closer to being a daily reality.

[15] These examples of renewable clean sources can be added to by others, such as geothermal and biomass power, both having their own benefits and applications.

[16] Geothermal power, which harnesses the natural heat below the Earth's surface, is used to heat homes or generate electricity. This resource is more effective in some regions than others. Iceland, for example, has a plentiful and easily reachable geothermal resource, while geothermal heat in the U.K., by comparison, is far less freely available.

[17] Biomass uses solid fuel created from plant materials to produce electricity. Although this energy source still requires the burning of organic materials, this is not wood and is now much cleaner and more energy efficient than in the past. Using agricultural, industrial, and domestic waste as solid, liquid and gas fuel is not only economical but also environmentally beneficial.

Can Clean Energy Replace Fossil Fuels?

[18] Humans have been using fossil fuels for decades, meaning that the switch to clean energy has been relatively recent. As a result, clean energy sources are still seen as being unpredictable and do not yet meet our global power demands. This means that clean energy is still being topped up with carbon-based energy sources.

[19] However, it is believed that our energy needs can be balanced by the efficient storing of clean energy so that it can be used when the demand is present. A great deal of work is being done to improve the infrastructures and storage capabilities of clean energy, with experts saying that clean and renewable energy could replace fossil fuels by 2050.

[20] Clean energy appears to be the future for the power needs of humanity across the globe as reliance on fossil fuels continues to **diminish**. As the **drive** towards clean, green, and renewable energy continues to advance, the cost will fall and work will

Unit 6　Clean Energy

be created to develop and install these new power solutions.

(1,154 words)

1. Judge whether the following statements are TRUE or FALSE.

(1) _____ Not all types of renewable energy are fully clean or green.

(2) _____ Clean energy can offset the financial and environmental expenses on imported fuels.

(3) _____ Renewable and clean energy industry is growing rapidly merely because it can create more job opportunities.

(4) _____ Using running water in municipal pipes as a form of energy has already become a reality.

(5) _____ The availability of geothermal power varies geographically.

2. Complete the following sentences with the words or phrases in the box below.

| topped up | scale up | incur | incentives |
| deforestation | replenish | harnesses | extracting |

(1) Governments at all levels are offering special tax _____ to micro, small and medium-sized enterprises.

(2) The lecture is very informative, introducing various ways to _____ the potential of texts, videos, and images to engage students in the learning process.

(3) Keep your phone between 50 and 80 percent charged, as depleting it completely or keeping it _____ all the time can decrease the battery's efficiency over time.

(4) Universities are encouraging greater efforts to _____ the use of the distance learning centers.

(5) The most common method of _____ shale oil is by surface mining, using technologies of pyrolysis, hydrogenation, or thermal dissolution.

(6) It's not advisable for college students to drop out of college to start their own business, for this may _____ the risk of extensive debt without getting a college degree.

(7) New teachers are urgently needed to _____ the teaching force as more teachers will retire over the next two years.

(8) Continuous practice of cutting trees is leading to _____, depriving animals and plants of their homes and grievously affecting the climate and species diversity.

3. Match the following technical terms with their Chinese equivalents.

I	II
(1) fuel spill	(　) 市政管道
(2) renewable energy	(　) 有机物质
(3) hydropower	(　) 地热能源
(4) natural gas leak	(　) 可再生能源
(5) municipal pipe	(　) 生物质能
(6) geothermal energy	(　) 水电
(7) biomass	(　) 天然气泄漏
(8) organic material	(　) 化石燃料
(9) carbon-based energy source	(　) 燃油泄漏
(10) fossil fuel	(　) 碳基能源

4. Translate the following paragraphs into Chinese.

Solar panels are frequently used for small electric tasks, such as charging batteries. While many people use solar energy for small garden lanterns, this same clean energy technology can be scaled up to larger panels that are used to provide power for homes or other buildings. Even multiple solar panels are arrayed and installed to power entire towns.

Clean energy appears to be the future for the power needs of humanity across the globe as reliance on fossil fuels continues to diminish. As the drive towards clean, green, and renewable energy continues to advance, the cost will fall and work will be created to develop and install these new power solutions.

Unit 6　Clean Energy

Phase III　Engaging Yourself

Warm-up Activity

Explain the following terms according to what you've explored before class.

♻ HFC　　♻ carbon emission quota　　♻ Global Energy Monitor

Further Reading

China's Green Transition: A Feasible Way of Quitting Coal

[1] Up to now, three revolutions have taken place in the energy industry. The first one took place 40,000 years ago, marked by the discovery and use of fire. The second one took place in the 18th century Britain, marked by large-scale use of coal. The third one began in the second half of the 19th century, marked by the use of electricity and internal engines, **ushering** in an era of energy utilization dominated by electricity and oil. Fossil energy resources represented by coal and oil promoted the second and third energy revolutions, and the rise of natural gas and new energy resources in recent years will lead to a new round of energy revolution. From coal to oil to natural gas and then to new energy resources, the trend of global energy utilization is reflected in the utilization of clean and low-carbon energy.

[2] As one of man's earliest sources of light and heat, coal has been the most cost-effective energy ever since its earliest commercial mining in 1750 near Richmond in Virginia, U.S. Meanwhile, fossil fuel-reliant power generation has led to a dramatic increase in CO_2 emissions since the industrial age began. CO_2 levels today are higher than at any point in the past 800,000 years. Studies show the last time the Earth saw such high CO_2 amounts was more than 3 million years ago. Even during the COVID-19 **pandemic** when electricity demand decreased due to **lockdown** and

133

production decline, average global CO_2 levels still hit a record high. If the trend continues, it will result in a four-time larger **cumulative** emission than the safe level set in the Paris Agreement. The UN's Intergovernmental Panel on Climate Change urged that countries worldwide need to **phase** out coal power by 2040 at the latest.

[3] What is China doing in reversing this trend and accelerating the global green transition?

[4] As the world's biggest **financier** of coal plants, China set its carbon-neutral goal in 2020, and just within one year, another step in the nation's green energy transition was taken. The Chinese government announced in September 2021 that it would stop building coal-fired power projects abroad.

[5] Currently, over 70 percent of the coal plants built globally rely on Chinese funding, and coal accounts for the largest share of China's overseas power sector investments. The country produces more than half of the world's coal-generated electricity. The policy change could affect 44 plants costing a **projected** $50 billion across 20 countries in Asia, Africa, South America, and Eastern Europe. Global Energy Monitor[1] (GEM) estimates that the decision could prevent 8 billion **tonnes** of additional carbon dioxide emissions over the lifetimes of the proposed plants.

[6] Between 2014 and 2020, only one of 52 overseas coal-fired power projects backed by Chinese funding went into operation, according to the International Institute of Green Finance, a Beijing-based think tank[2]. Chinese-financed coal plants worth more than $65 billion have either been canceled or **mothballed**. China's move would save several countries from pouring billions of dollars into coal plants that would quickly become **stranded** assets[3], given the declining costs of renewables and public **momentum** for carbon restrictions.

1　Global Energy Monitor：全球能源监测组织

2　think tank：（政治、社会、经济问题的）智囊团；智库；专家小组

3　stranded asset：搁浅资产（指由于环境、政策或市场变化等原因导致价值大幅下降或无法实现预期收益的资产）

Unit 6　Clean Energy

[7] In the past few years, the investment for new coal plants has shrunk globally. Countries which heavily depend on coal are facing an urgent need to adjust their energy supply chain and expand renewable power generation. Bangladesh, where around 35 million people (some 30 percent of its population) are living in coastal areas threatened by climate change, has canceled 10 proposed coal-fired power projects, and the office of the President of Pakistan has become the world's first President's House fully powered by renewable energy. Pakistan, one of the key countries of China's Belt and Road Initiative (BRI), would also stop building new coal plants.

[8] As more countries are **pledging** to abandon coal, power giants are **eyeing** on clean energy sector, which will create millions of new jobs. Roadmap to Net-Zero by 2050 Report of the International Energy Agency shows that during the transition to net-zero emissions, over 30 million new jobs related to clean energy and low-emissions technologies will be generated by 2030.

[9] Over the past decade, Chinese investment in overseas renewable energy projects has risen steadily. In 2020, Chinese investment overseas in solar, wind, and hydropower projects surpassed that in fossil fuels for the first time, following the launch of the BRI in 2013. Meanwhile, China also promised to better support developing countries in utilizing green and low-carbon energy. In September 2021, China began to enforce the Kigali Amendment to the Montreal Protocol to curb the emission of hydrofluorocarbons (HFCs), a greenhouse gas that is over 14,000 times more powerful than CO_2 in warming the planet. If fully implemented, it will prevent more than 0.4 °C of global warming by the end of this century.

[10] However, concerns remain over how China will tackle domestic emissions. The country has been heavily reliant on coal energy for building a vast amount of infrastructure. Coal consumption still makes up over half of the country's total energy demand despite a 70-percent decrease over the past decade.

[11] In October 2021, the government issued a guideline on green development

aimed at cutting CO_2 emissions in both urban and rural areas. A white paper released by the State Council Information Office targets a steep cut in carbon intensity and a 5-percent increase in the share of non-fossil fuels in primary energy consumption. China has diversified its energy supply structure and gradually replaced some coal use with cleaner-burning fuels. Apart from a rapidly expanding capacity of solar and wind, other renewable energy sources like hydroelectric sources (8 percent), natural gas (8 percent), and nuclear power (2 percent) accounted for a small but growing proportion of China's energy consumption.

[12] China, the world's top emitter of greenhouse gases, has been **piloting** emissions trading since 2011 in seven provinces and municipalities, including Beijing, Shanghai, and Guangdong, to pave the way for the establishment of a national Emissions Trading System (ETS). The pilot programs have covered key emission industries like steel and cement and have driven major emitters to reduce their emissions. In July 2021, China set its first national ETS to encourage power companies to cut down carbon emissions. More corporate income tax credits[1] would be added to support environmental protection, and authorities would also give more financial support to green and low-carbon development.

[13] Prior to July 2021, a set of interim rules for carbon emissions trading management in China came into effect, a major effort following the establishment of the ETS. A total of 2,225 power firms across the country assigned with CO_2 emission caps can trade their emission **quotas** via the system.

[14] Still, reaching the carbon-neutral goal requires time, as well as a feasible way to balance the speed of going green and maintaining energy security. More importantly, the pace of **flattening** carbon emissions depends on how fast countries can **tap** renewable energy resources and phase out coal use **coherently**.

(1,152 words)

---

1 income tax credit: 所得税抵免（一种减轻纳税人税收负担的政策，允许纳税人在计算应纳税额时抵扣一定数额的所得税）

Unit 6　Clean Energy

1. **International Institute of Green Finance 国际绿色金融研究院**

   清华大学设立的研究机构，致力于推动绿色金融领域的研究和实践。通过开展科学研究、政策分析、人才培养等活动，为绿色金融发展提供智力支持和专业指导。

2. **International Energy Agency 国际能源署**

   简称 IEA，成立于 1974 年，总部位于巴黎。主要任务包括收集和分析全球能源数据、制定政策建议以应对能源挑战、促进世界范围内的能源安全、可持续能源发展和协调国际合作。

3. **Kigali Amendment to the ontreal Protocol《蒙特利尔议定书基加利修正案》**

   国际环保领域的一项法律文书，于 2016 年在卢旺达的基加利通过。该修正案旨在加强《蒙特利尔议定书》中对氢氟碳化物（HFCs）的控制，逐步减少和最终消除 HFCs 的生产和使用，减缓气候变化和保护地球环境。该修正案的通过被视为环保领域的重大里程碑，表明了国际社会对减少温室气体排放和实现可持续发展的承诺和合作精神。

1. Read the text and choose the best answer.

(1) What can we learn from the text?

　A. The history of global energy use is characterized by the utilization of cleaner and low-carbon energy.

　B. Fossil fuels are the primary cause of increase in CO_2 emissions throughout human history.

　C. CO_2 levels were lower than today as long as 800,000 years ago.

　D. CO_2 emissions are expected to reach zero since 2040 at the latest.

(2) According to the text, which of the following cleaner-burning fuels has the largest proportion of consumption in China?

   A. Hydroelectric sources.

   B. Natural gas.

   C. Nuclear power.

   D. Solar and wind sources.

(3) What is the Chinese government's purpose of setting ETS?

   A. To add more income tax credits.

   B. To give more financial support to low-carbon development.

   C. To encourage CO_2 emitters to cut down on their carbon emissions.

   D. To pilot emissions trading.

(4) Which of the following exemplifies China's efforts in green energy transition?

   A. The Chinese government promised to stop building coal-fired power projects abroad.

   B. Only one of 52 overseas Chinese-funded coal-fired power projects went into operation between 2014 and 2020.

   C. Chinese investment overseas in cleaner energy projects surpassed that in fossil fuels for the first time in 2020.

   D. All of the above.

(5) Which of the following does **NOT** directly affect the goal of carbon neutrality?

   A. The balance between green energy use and energy security.

   B. The emission quotas that can be traded in ETS.

   C. The pace of phasing out coal use.

   D. The speed of exploiting renewable energy resources.

2. **Replace the underlined parts with the words or phrases in the box below.**

| cumulative | projected | stranded | mothball |
| tap | flatten | phased out | think tank |

Unit 6   Clean Energy

(1) Efforts to reduce energy consumption have helped <u>level out</u> the curve of carbon emissions in recent years.

(2) Whether to <u>halt</u> the construction of the paper mill has not been decided yet, because the opinions of the parties concerned are hugely far apart.

(3) By the end of last year, the <u>total</u> installed capacity of new energy storage projects completed and put into operation in China had exceeded 17 million kilowatts.

(4) Chipmakers in the U.S. are <u>expected</u> to add about 115,000 jobs by 2030, but 58 percent of these jobs could remain unfilled because of a significant shortage of semiconductor talent.

(5) The cruise ship is a project sponsored by the Ministry of Industry and Information Technology and the Ministry of Transport in a bid to <u>harness</u> clean energy and green transportation.

(6) By the end of 2022, China had <u>eliminated</u> a total of nearly 300 million tons of steel with outdated and excess production capacity, according to Ministry of Ecology and Environment.

(7) The founder of the newly established global <u>think factory</u> headquartered in Islamabad is an international expert specializing in the Belt and Road Initiative, Afghanistan, South and Central Asia.

(8) Caught in a snowstorm during the climb, they were <u>trapped</u> half-way up the mountain for nearly three hours before finally rescued.

3. Write a summary of the text in 120–150 words. You may use the given words or expressions in the box.

announce	carbon emissions
move	declining costs
Chinese investment	fossil fuels
pledge	domestic emissions
coal consumption	Emissions Trading System

139

Unit 6　Clean Energy

Phase IV　Commitment

1. Translate the following paragraph into English.

　　新型储能是指以锂电池为代表的、电化学储能为主体的储能方式。新型储能技术具有良好的安全记录、高性价比和环境友好性，适合大规模储能应用。中国完善的锂电池产业链以及压缩空气储能和液流电池（flow batteries）等技术应用范围的逐步扩大，将为新能源存储行业带来一段强劲的增长期。据国家能源局（National Energy Administration）的数据，截至 2023 年 6 月底，中国新建成并投产的新型储能项目累计装机容量已超过 17.33 千兆瓦。

2. Work in pairs to discuss the following topics.

(1) Nuclear energy and solar energy are two forms of energy sources. Which do you think has the potential for providing most of the energy needed by humanity in the near future?

(2) Fueling automotive vehicles in the future will be a major challenge. Which technology is more promising, batteries or synthetic fuels produced with solar energy? Why?

3. Work in groups to make a presentation on any topic related to the theme of the unit.

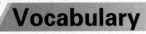

Vocabulary

Active Reading: All We Need to Know About Clean Energy

crossover	/ˈkrɒsəʊvə(r)/	*n.*	a point or place of crossing from one side to another	横跨处；交叉
hydropower	/ˈhaɪdrəʊpaʊə(r)/	*n.*	electricity produced using the power of water	水力发电
spill	/spɪl/	*n.*	an amount of sth. that has come out of a container	洒出（量）；泼出（量）；溢出（量）；泄漏（量）
incur	/ɪnˈkɜː(r)/	*v.*	to experience sth., usually sth. unpleasant, as a result of actions you have taken	招致；遭受
windmill	/ˈwɪndˌmɪl/	*n.*	a building or structure with parts that turn around in the wind, used for producing electrical power or crushing grain	风车房；风力磨坊；风车
grind	/ɡraɪnd/	*v.*	to break sth. such as corn or coffee beans into small pieces or powder, either in a machine or between two hard surfaces	磨碎；碾碎；把……磨成粉
windfarm	/ˈwɪndfɑːm/	*n.*	a place where a lot of windmills or wind turbines have been built in order to produce electricity	风力发电场
municipal	/mjuːˈnɪsəpəl/	*adj.*	relating to or belonging to the government of a town or city	市政的；市立的

Unit 6　Clean Energy

harness	/'hɑːnəs/	v.	to control and use the natural force or power of sth.	控制；利用（以产生能量等）
diminish	/dɪ'mɪnɪʃ/	v.	to become or make sth. become smaller, weaker or less important	减少；减小；降低
drive	/draɪv/	n.	an effort to achieve sth., especially an effort by an organization for a particular purpose	（团体为达到某目的而进行的）有组织的努力；运动

Further Reading: China's Green Transition: A Feasible Way of Quitting Coal

usher	/'ʌʃə(r)/	v.	to show someone where he or she should go, or to make someone go where you want him or her to go	引导；引领；把……引往
pandemic	/pæn'demɪk/	n.	a disease that affects people over a very large area or the whole world	大规模流行的疾病；广泛蔓延的疾病
lockdown	/'lɒkdaʊn/	n.	an official order to control the movement of people or vehicles because of a dangerous situation	活动（或行动）限制
cumulative	/'kjuːmjələtɪv/	adj.	increasing gradually as more of sth. is added or happens	累积的；渐增的
phase	/feɪs/	v.	to make sth. happen gradually in a planned way; to arrange to do sth. gradually in stages over a period of time	分阶段进行；逐步做

143

financier	/faɪˈnænsɪə(r)/	*n.*	a person who manages or lends large amounts of money for or to businesses or governments	金融家；投资家；（为项目或企业）提供资金者
project	/prəˈdʒekt/	*v.*	to calculate what sth. will be in the future, using the information already known	预计；推算
tonne	/tʌn/	*n.*	a unit for measuring weight, equal to 1,000 kilograms	公吨（等于1 000公斤）
mothball	/ˈmɒθbɔːl/	*v.*	to stop using a factory, equipment etc. or to not continue with a plan, temporarily but possibly for a long time	封存；保藏；把……束之高阁；搁置不用
strand	/strænd/	*v.*	to leave in a strange or an unfavorable place, especially without funds or means to depart	使滞留
asset	/ˈæset/	*n.*	sth. valuable belonging to a person or an organization that can be used for the payment of debts	资产；财产
momentum	/məʊˈmentəm/	*n.*	the force that keeps an object moving or keeps an event developing after it has started	动量；冲量；冲力；推动力；势头
pledge	/pledʒ/	*v.*	to make a formal, usually. public, promise to give or do sth.	保证给予（或做）；正式承诺
eye	/aɪ/	*v.*	to watch or look at sth. or sb. in a very close or careful way	注视；审视；细看

Unit 6 Clean Energy

pilot	/ˈpaɪlət/	*v.*	to test a new idea, product, etc. on people to find out whether it will be successful	试点；试行
quota	/ˈkwəʊtə/	*n.*	an official limit on the number or amount of sth. that is allowed in a particular period	定额；限额；配额
flatten	/ˈflætən/	*v.*	to make or become level or less steep	使变得平坦；使降低到一定水平
tap	/tæp/	*v.*	to make use of a source of energy, knowledge, etc. that already exists	利用；开发；发掘（已有的资源、知识等）
coherently	/kəʊˈhɪərəntli/	*adv.*	(working) closely and well together	紧密协作地

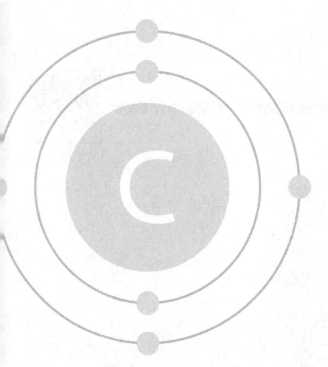
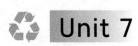

Unit 7

Transport Decarbonization

If the rate of change on the outside is greater than the rate of change on the inside, the end is near.

—Jack Welch

Phase I First Sight

Video 1 The Surprisingly Long History of Electric Cars

New Words and Phrases

boiler /ˈbɔɪlə(r)/ *n.* 汽锅；锅炉

lug /lʌg/ *v.* 用力拉或拖

crank /kræŋk/ *v.* 用曲柄转动（或启动）

foul /faʊl/ *adj.* 难闻的；令人恶心的

swap out 换出；置换出

luxury /ˈlʌkʃəri/ *adj.* 奢侈的；豪华的

muffler /ˈmʌflə(r)/ *n.* 消音器

engine mount 发动机支架

vibration /vaɪˈbreɪʃən/ *n.* 震动；颤动；抖动

smog /smɒg/ *n.* 烟雾；雾霾

portable /ˈpɔːtəbəl/ *adj.* 便携式的；轻便的

nickel metal hydride battery 镍氢电池

mandate /ˈmændeɪt/ *n.* 命令；指令

roadster /ˈrəʊdstə(r)/ *n.* 跑车；双人座敞篷汽车

upfront /ˌʌpˈfrʌnt/ *adj.* 预付的；预交的

aggressive /əˈgresɪv/ *adj.* 声势浩大的

reclaim /rɪˈkleɪm/ *v.* 取回，拿回

rearview /ˈrɪəvjuː/ *n.* 后视镜

Unit 7 Transport Decarbonization

1. Watch the video and choose the best answer.

(1) If you drove a steam-powered car over a long distance in 1899, what would you have to do?

   A. To have several boilers for power.

   B. To carry extra water to refuel.

   C. To drive with the heated engine.

   D. To drive as far as I could.

(2) Which is **NOT** the correct description of the 19th-century electric vehicles?

   A. Quiet to run.

   B. Easy to refuel.

   C. Dangerous to drive.

   D. Quick to start.

(3) What was the major problem with the early electric vehicles?

   A. It was difficult to operate an electric vehicle.

   B. Many people could not afford electric vehicles.

   C. The government did not encourage electric vehicles.

   D. The electric vehicles were not appealing to drivers' taste.

(4) Why did gas-powered cars win over electric cars in the 1910s?

   A. Because Ford's Model T captured the public attention.

   B. Because automakers helped with easy access to low priced oil.

   C. Because the sales of electric cars dropped sharply for their inefficiency.

   D. Because the qualities of the internal combustion engines were improved.

(5) What sparked a new wave of electric vehicles in the 1980s?

   A. Automakers developed efficient batteries from new materials.

   B. Governmental regulations required the reduction of smog.

   C. Automakers slowed investments in gas-powered vehicles.

   D. Both A and B.

2. Watch the video again and complete the sentences.

(1) The internal combustion engines in these models required dangerous hand-cranking to start and emitted loud noises and foul-smelling _____ while driving.

(2) By the end of the 19th century, battery-powered electric cars were a popular and reliable _____ to their occasionally explosive competitors.

(3) In the 1980s, car companies had spent decades investing in internal combustion engines without devoting any resources to solving the _____ battery problem.

(4) Hybrid cars aren't true electric vehicles; their nickel metal hydride batteries are only used to _____ the efficiency of gas-burning engines.

(5) But in 2008, Tesla Motors went further, grabbing the attention of consumers, automakers, and _____ with its lithium-ion-powered roadster.

(6) Electric vehicles are now expected to replace gas-powered ones altogether. Soon, electric cars will reclaim their place on the road, putting gasoline in our _____.

3. Answer the following questions according to the video.

(1) How do you evaluate steam-powered cars, gasoline-powered cars, and battery-powered electric cars respectively?

(2) Do you agree with the statement in the video that electric cars will reclaim their place on the road? Why or why not?

Video II Can Flying Go Green?

New Words and Phrases

fuel cell 燃料电池

aviation /ˌeɪviˈeɪʃən/ *n.* 航空工业；飞机工业

front runner 领先者

Unit 7   Transport Decarbonization

long-haul flight 长途飞行

sophisticated /səˈfɪstɪkeɪtɪd/ *adj.* 复杂巧妙的；先进的

biofuel /ˈbaɪəʊfjuːəl/ *n.* 生物燃料

synthetic fuel 合成燃料

algae /ˈældʒiː/ *n.* 水藻；海藻

pump up 打气；增加

roll out 推出；推广；展开

prohibitively /prəˈhɪbətɪvli/ *adv.* 过高地；过分地

stump up 支付（一笔钱，尤指不情愿地支付）

premium /ˈpriːmiəm/ *n.*（正常价格或费用以外的）加付款；加价

1. **Watch the video and judge whether the following statements are TRUE or FALSE.**

(1) _____ The EU holds the belief that by 2060 hydrogen technologies could reduce the industry's carbon emissions by up to 70 percent.

(2) _____ Hydrogen packs more energy into a given volume than jet fuel. So powering long-haul flights will mean extensive aircraft redesign.

(3) _____ The big problem for aviation is the pricy green technologies in the face of the worst economic crisis.

(4) _____ One of the benefits of going down the sustainable aviation fuel, the biofuel, is a drop in profits and price.

(5) _____ Aviation companies would love to buy sustainable aviation fuel, if it was only 50 percent more expensive than conventional jet fuel.

2. **Watch the video again and fill in the blanks with the words you hear.**

Without decarbonizing technologies like hydrogen, aviation could be responsible for 22 percent of the planet's emissions by 2050. Because it can (1) _____ more

151

energy into a smaller space, hydrogen has recently (2) _____ electric batteries as the front runner in the race for cleaner flying.

But there are plenty of (3) _____ to overcome before these hydrogen technologies are (4) _____ viable, such as infrastructure, renewable energy, and the size of the energy tank.

Even with the government's help, it will take at least two (5) _____ before hydrogen-powered planes are ready for takeoff. To reduce emissions in the nearer term, aviation needs other (6) _____. And the front runner has recently emerged.

Going green may be the biggest challenge the aviation industry has ever faced. It demands the right (7) _____ and policies to support a range of technologies, and that might mean passengers need (8) _____ pockets.

3. Answer the following questions according to the video.

(1) What are the problems with the present aviation industry?
(2) What stakeholders might be involved in the program of driving the aviation industry to go green?

Unit 7 Transport Decarbonization

Phase II Getting to Know

Warm-up Activity

Explain the following terms according to what you've explored before class.

- fuel-cell vehicle (FCV)
- hybrid car
- zero-emissions zone (ZEZ)
- black carbon

Active Reading

A Strategy to Decarbonize the Global Transport Sector

Introduction

[1] Global demand for passenger and freight transportation is trending strongly upward, driven by population and economic growth, and in particular by a rapidly expanding **cohort** of new middle-class consumers. As transportation demand has grown, so too, **inexorably**, have carbon emissions from the global transportation sector. That is a trend that cannot be permitted to continue. The destructive effects of the warming that has already occurred as a consequence of **anthropogenic** emissions of carbon dioxide and other greenhouse gases are **dire** enough. Scientists warn that we must steeply reduce greenhouse gas emissions by mid-century to avoid additional warming that will have genuinely catastrophic effects. In that light, the task of transforming the technologies and systems that move people and goods around the world seems imperative, however **daunting**.

[2] And greenhouse gas emissions do not even fully describe the environmental challenges posed by rising demand for transportation in a system that remains dependent primarily, or even entirely, on burning fossil fuel. The public health toll from air pollution, especially in many large cities, remains unacceptably high. Climate

and health impacts are inherently **coupled**. For example, particulate matter (PM) in diesel engine exhaust has significant and direct **adverse** impacts on both health and climate, and electric vehicles with zero tailpipe emissions bring significant co-benefits in improved air quality.

[3] The bottom line is that a sector that is almost **exclusively** dependent on a single energy source, petroleum, operating on infrastructure that represents trillions of dollars of investment over many decades, must change **substantially** in little more than a generation. Efforts to limit pollution and climate impacts of existing technologies must be continued and strengthened. Yet that will not be enough. Ambitious policies, incentives, and investments to bring forth new, clean transportation technologies and systems must be put in place without delay.

What Is the Baseline?

[4] The transportation sector accounts for approximately one-quarter of global anthropogenic CO_2 emissions. Global CO_2 emissions from transportation will grow significantly over the next 30 years, from approximately 12 Gt in 2020 to 21 Gt annually in 2050. Two-thirds of transportation CO_2 emissions in 2020 were from the four largest vehicle markets and the marine and aviation sectors. In terms of the contribution of different modes of transportation, on-road vehicles dominated, accounting for roughly 77 percent of global transportation CO_2 emissions in 2020 (43 percent from light-duty vehicles including 2 and 3 wheelers and 34 percent from heavy-duty vehicles including buses). Heavy-duty vehicles made a **disproportionate** contribution to climate and air pollution relative to their numbers in the global vehicle fleet, in part because of their substantial emissions of particulate matter, including black carbon, which has high short-term warming potential.

[5] Public health impacts from transportation emissions have continued to rise despite progress on reducing emissions per vehicle-kilometer-traveled (VKT). It is estimated that exposure to transportation emissions resulted in 7.8 million years of life lost and approximately $1 trillion in health damages globally in 2015. Despite the

adoption of more **stringent** vehicle regulations in many countries, the transportation sector remains a significant contributor to the global burden of air pollution-related **morbidity** and **mortality** which is especially concentrated in certain countries and regions.

What Is the Target?

[6] Achieving the Paris Agreement objectives to limit the global temperature increase to well below 2 °C this century and pursue efforts to limit the temperature increase even further to 1.5 °C, will require steep reductions in carbon emissions from all sectors. Although the existing and emerging policies and technologies can generate some reductions, they come close but do not fully achieve the magnitude of reductions required to enable the transport sector to fully contribute to keeping global warming below 1.5 °C. Besides, additional reductions will be required from policies that reduce vehicle miles traveled (e.g., increases in public transit ridership, increased rates of walking and biking, compact development, and so on) to achieve our 2050 target of 2.6 Gt of CO_2.

What Is an Ambitious Yet Feasible Scenario?

[7] Achieving the global CO_2 target for the transportation sector will mean the accelerated **deployment** of existing and emerging low-emission and zero-emissions technologies across all transportation segments. One of the key focus areas is to track technology developments across transportation segments: light-duty vehicles, heavy-duty vehicles, marine, aviation. With this baseline knowledge, we are well placed to propose an ambitious yet feasible decarbonization scenario: technologically feasible while also taking into account such practical considerations as cost, time required to achieve large-scale production and deployment, and differences in advertised versus real-world emissions. Even under the most optimistic decarbonization scenarios, more than two billion new internal combustion engine vehicles will be sold over the next 30 years. It is critical that these vehicles be as efficient as possible. If the transportation sector is to be decarbonized, ultimately the vast majority of vehicles

must produce zero tailpipe emissions and be fueled by carbon-neutral and renewable energy sources. Focus can be put on technological improvements across transportation segments as well as **upstream** fuel and electricity production.

What Is the Highest Priority Work over the Next Five Years?

[8] Summarized below are the most important elements of the strategic approaches to electric and conventional vehicles over the next five years in the hope of benefiting other **stakeholders** as well.

- Electric vehicles are the single most important technology for decarbonizing the transport sector.
- Continued progress on energy efficiency is critical to achieving our climate goals and supporting the transition to **electrification**.
- Reduction of black carbon through the introduction of world-class standards for vehicles and fuels in as many markets as possible by 2025 is a **synergistic** strategy pursuing both health and climate benefits.
- Research and development targeted at zero-emissions technologies are needed for the most challenging segments, such as international marine and aviation.
- The scale of the crisis we are facing **necessitates** increased collaboration and communication.

Concluding Thoughts

[9] Our aim is to support the achievement of the Paris Climate Agreement targets to limit the global temperature increase this century to well below 2 °C and to pursue efforts to limit the temperature increase to 1.5 °C by reducing annual global transportation CO_2 emissions to an estimated 2.6 billion metric tons, as a **median** target, in 2050. We believe that the key to achieving this target is through the deployment of stringent regulations that promote the rapid adoption of new low and zero-emissions technologies in the transportation sector and the electricity grid. Moving forward it is important to develop and implement more detailed and targeted strategies for each region or market in order to ensure that relevant policymaking is

Unit 7 Transport Decarbonization

well supported. There are key metrics that can be tracked in the short term in order to show progress. These metrics include: adoption and implementation of relevant policies; conventional and electric vehicle sales; global sales of transportation fuel; real-world emissions to measure compliance with regulations; and investment in development and production of new technologies. Tracking metrics such as these on an annual basis will not only give us insight into the progress being made towards the longer-term target, but will inform any necessary strategy adjustments.

(1,156 words)

1. Judge whether the following statements are TRUE or FALSE.

(1) _____ Heavy-duty vehicles emit black carbon that constitutes the majority of the global transportation greenhouse gas and CO_2.

(2) _____ Transportation contributes significantly to air pollution-related diseases and deaths.

(3) _____ In addition to the existing policies of technologies, more is required of policies to reduce vehicle miles traveled.

(4) _____ A feasible roadmap to the global CO_2 target is to deploy decarbonization technologies at all cost without delay.

(5) _____ The key to achieving global climate goals is continuing progress in energy efficiency.

2. Complete the following sentences with the words or phrases in the box below.

adverse	coupled	deploy	exclusively
stringent	inexorably	magnitude	put in place

(1) The spokesman said an ongoing enhanced quality assurance process had also been _____ to ensure the potential objective.

(2) Advancing technologies and strict laws are helping control some of these _____ environmental effects.

(3) Climate change cannot be battled by relying _____ on cutting emissions or market-based solutions.

(4) Overproduction, _____ with falling sales, has led to huge losses for the company.

(5) In some cases, _____ job dismissal regulations adversely affect productivity growth and hamper both prosperity and overall well-being.

(6) The amendment makes it a crime for anyone to _____ peer-to-peer technology that facilitates the exchange of copyrighted material online.

(7) The two countries say they are aware of the _____ of the problem, and are taking steps to slow the growth of emissions.

(8) The background level of tropospheric ozone is _____ rising as volatile organic compounds are released from burning coal and oil.

3. Match the following technical terms with their Chinese equivalents.

I	II
(1) heavy-duty vehicle ( )	零废气排放
(2) internal combustion engine ( )	柴油发动机的废气
(3) black carbon ( )	电气化
(4) zero tailpipe emission ( )	重型车辆
(5) diesel engine exhaust ( )	新能源汽车
(6) public transit ridership ( )	车辆行驶里程
(7) electrification ( )	黑碳
(8) vehicle fleet ( )	内燃机
(9) new-energy vehicle (NEV) ( )	车队
(10) vehicle-kilometer-traveled ( )	公共交通客运量

4. Translate the following paragraph into Chinese.

The key to achieving global climate target is through the deployment of stringent regulations that promote the rapid adoption of new low and zero-emissions technologies in the transportation sector and the electricity grid. Moving forward it is important to develop and implement more detailed and targeted strategies for each

Unit 7 Transport Decarbonization

region or market in order to ensure that relevant policymaking is well supported. There are key metrics that can be tracked in the short term in order to show progress. These metrics include: adoption and implementation of relevant policies; conventional and electric vehicle sales; global sales of transportation fuel; real-world emissions to measure compliance with regulations; and investment in development and production of new technologies.

Phase III Engaging Yourself

Warm-up Activity

Explain the following terms according to what you've explored before class.

- synfuel
- particulate matter
- carbon-intensive industry

Further Reading

How Can Airlines Chart a Path to Zero-Carbon Flying?

[1] The airline industry is understandably likely to undergo structural changes with regard to demand and the degree of industry consolidation, the health and livelihoods of its millions of workers, along with unprecedented government support. That transition provides an opportunity to rebuild the industry for a low-carbon future, something that airlines have been **grappling** with for some time to face the challenge of reducing its carbon emissions in line with international goals to reach net-zero emissions by 2050.

[2] Airlines are working to align emissions cuts with their bottom-line interests. They have encouraged operational efficiency and optimal air-traffic management and invested billions of dollars to modernize aircraft with more efficient **aerodynamics** and engines using lighter-weight materials. However, these actions get the industry only so far, cutting emissions by no more than 20 to 30 percent compared with the do-nothing alternative.

Operational Efficiency

[3] Fuel typically accounts for 20 to 30 percent of operational costs — one of the largest single cost items. Airlines therefore have an **intrinsic** motivation for adopting more fuel-efficient flying, **taxiing**, and airport operations. They are also **eking** out

fuel-efficiency gains by decreasing the extra fuel loaded onto aircraft and introducing lighter materials to reduce aircraft weight.

[4] In a recent survey of airlines, despite these efficiency gains, carriers capture only around 50 percent of their full potential. Only a few airlines address their employees' behavior and mindsets related to fuel. This is a crucial area, since pilots, **dispatchers**, and other airline employees have considerable **discretion** in preparing and conducting safe flights, with direct implications for fuel consumption. To increase fuel efficiency, airlines should identify the areas needing improvement with the help of analytics and systematically drive behavioral change with their frontline employees.

[5] Airlines also consume additional fuel from **zigzagging** through nations' ATM sectors that require predefined handovers. Other inefficiencies include limits on air-traffic-control capacity and a lack of automation in air-navigation services. Eliminating those inefficiencies requires a joint effort from a large group of stakeholders, including governments, regulators, and militaries, which makes the process painfully slow.

New Aircraft Technology

[6] Airlines invested almost $120 billion in new aircraft in 2018 alone. New models have highly efficient engines, and modern long-haul twin-engine aircraft are replacing four-engine aircraft, which enables up to 20 percent fuel-efficiency improvement per passenger. Regarding commercial-fleet strategy, executives should consider not just fuel-price predictions but also the future cost of carbon. Applying carbon emissions as a fuel-cost premium could lead to an accelerated fleet rollover and faster adaption of future aircraft technology, including some electrification.

[7] Alternative **propulsion** (such as via electricity and hydrogen) could one day replace conventional turbine-powered planes, especially smaller aircraft on shorter flights. However, the use of fully electric aircraft carrying more than 100 passengers appears unlikely within the next 30 years or longer for the sake of battery weight. Electric propulsion could start with hybrid or turboelectric flying, enabling further improvements in fuel efficiency as jet engines become smaller and lighter, using less fuel.

[8] Aircraft could also be powered by hydrogen, either from direct combustion (hydrogen turbine) or via a fuel cell. Hydrogen emits no CO_2 during the combustion process and allows for significant reduction of other elements that drive global warming, such as **soot**, nitrogen oxides, and high-altitude water vapor. While smaller hydrogen-powered aircraft could become feasible in the next decade, aircraft with more than approximately 100 passengers should develop significant aircraft-technologies, and infrastructure constraints on refueling would need to be overcome.

Sustainable Aviation Fuel (SAF)

[9] SAF is a solution that can achieve full decarbonization, but it comes with challenges on both the supply and demand fronts.

[10] Use of advanced biofuels is a likely near-term solution. The technical feasibility of fuel made from vegetable or waste oils is proven, the product is certified, and some airlines use the fuel in daily operations. But getting the appropriate **feedstock** and supply chain in place is difficult; building production facilities and **refineries** is costly. Used cooking oil, a popular ingredient for biofuel, has **fragmented** availability and is expensive to collect. Other vegetable oils have high costs of production, collection, transportation, and conversion to fuel.

[11] Feedstock resources also involve other environmental risks, such as deforestation and the creation of **monocultures**. Feedstock sources for biofuels must be selected thoughtfully to limit food versus fuel challenges. Some airlines, including Cathay Pacific Airways and United Airlines, have invested in facilities to demonstrate how municipal household waste could be **gasified** and subsequently turned into jet fuel. In some regions, the **fermentation** of wood **residues** into sustainable kerosene has shown potential as a viable path.

[12] Alternatively, the use of synfuels derived from hydrogen and captured carbon emissions could become a scalable option. Such synfuels require water, renewable electricity to produce hydrogen, and CO_2. Today, these power-to-liquid fuels are several times the cost of conventional kerosene, though we expect a significant cost reduction for green hydrogen (via reduced costs of renewable electricity and

electrolyzers) in the coming years. In the first step, CO_2 could be captured as waste gas from carbon-intensive industries, such as steel, chemicals, and cement. In the long term, to become net-zero CO_2, the required CO_2 needs to be extracted from the carbon cycle (taken from the air with direct air capture). While this is costly today, the process benefits from cheaper renewable-electricity generation in the future.

[13] While synfuels could become an answer to cutting emissions over the long run, it is unclear, at this point, which SAF sources will emerge as winners. In effect, SAF presents a classic chicken-and-egg problem. Airlines don't yet have a viable business case for buying SAF; therefore, its production volume is small, with little economies of scale and insufficient funding.

More Stakeholders for SAF

[14] Breaking through the which-comes-first problem with SAF would involve a number of groups, each doing its part to put the puzzle together. First, airlines could build and **orchestrate** a **consortium** of stakeholders that includes technology providers and oil companies to drive demand and help bridge the cost gap. For example, airlines could commit to buying SAF at a predefined price, or at a price differential to traditional jet fuel, which would eliminate market risks for fuel suppliers.

[15] Second, financial institutions could provide venture capital for building SAF-production facilities and new infrastructure that allows for the anticipated cost savings. Building a **coalition** of airlines could increase the required volume, resulting in scale effects.

[16] Third, airlines could work with customers willing to pay a premium for the opportunity to decarbonize their employees' footprints. Microsoft was committed to reducing its environmental footprint by promoting SAF and paying for the cost premium. For individual customers, airlines could use loyalty-program rewards as incentives to offset CO_2 through SAF use.

[17] Fourth, policymakers at domestic and regional levels could play a critical role by creating incentives for SAF production and setting appropriate targets.

Countries such as Canada and Norway that are willing to apply blending mandates are moving forward on this front. Policymakers could also reallocate aviation taxes back to the industry to fund decarbonization, closing the remaining cost gap between conventional kerosene and SAF.

[18] As the aviation industry emerges from the painful period, there is an opportunity to move closer to low-carbon goals. The aviation industry has made great strides in fuel efficiency and operational advancements. But to reach global emission-reduction targets, it will need to move to the next level of decarbonization, and SAF is an option that could get it there. Bolder moves and much deeper **collaboration** among stakeholders are necessary to build financial structures and programs that can help **funnel** capital into SAF production.

(1,255 words)

1. Read the text and choose the best answer.

(1) Aviation is likely to undergo structural transition because of _____.

   A. the demand and degree of industry consolidation

   B. the demand of the health and livelihoods of its workers

   C. the challenge of global climate goals

   D. all of the above

(2) Which of the following is **NOT** mentioned as a factor affecting aviation operational efficiency?

   A. Fuel cost.

   B. The behavior of airline employees.

   C. The mindsets of passengers.

   D. Air-traffic-control capacity.

(3) For the time being, fully electric aircraft carrying more than 100 passengers appears unlikely because of _____.

   A. water supply

   B. battery weight

C. turbine efficiency

D. infrastructure constraints

(4) What has prevented synfuels from becoming a scalable option for fuel?

A. High cost of hydrogen production.

B. Carbon exhaustion.

C. Electricity generation.

D. Heavy water weight.

(5) According to the passage, which plays a critical role in aviation decarbonization?

A. Individual airlines.

B. Financial institutions.

C. Policymakers.

D. Oil companies.

2. Replace the underlined parts with the words or phrases in the box below.

| align | discretion | eke out | feasible |
| funnel | grapple with | intrinsic | orchestrate |

(1) The new government has yet to deal with the problem of air pollution.

(2) Sustainability is a useful template to match short-term policies with medium-to-long-term goals.

(3) The strategic petroleum reserve faces inherent physical constraints in supplying the oil market.

(4) Most city refugees have to manage an existence in illegal or informal sectors of the economy.

(5) Technological improvements are needed so that wind, solar, and hydrogen can become more viable parts of the energy equation.

(6) It helps organize the assets of a business to form highly optimized and effective processes.

(7) The solution for pensioners is to pour about half of their payroll taxes into personal retirement accounts.

(8) How much to tell terminally ill patients is left to the <u>caution</u> of the doctor.

3. Write a summary of the text in 120–150 words. You may use the given words in the box.

align	decarbonization	transition	challenge
collaborate	optimal	undergo	stakeholder

Unit 7 Transport Decarbonization

Phase IV Commitment

1. Translate the following paragraph into English.

  在优厚的补贴和惊人的销售业绩驱动下，已有200多家公司宣布有意在中国生产和销售新能源汽车。中国政府正在推动新能源汽车（部分或完全电动的汽车）的发展，将其作为《中国制造2025》这一新产业政策的一部分，希望届时在十个高科技行业缔造国家冠军企业。但对有些企业而言，这股淘金潮尚未开始就可能结束了。中国正在淘汰规模较小的新能源汽车公司，因为其中有些公司虽消耗了政府补贴，却没有生产出可商用的汽车。

2. Work in pairs to discuss the following topics.

(1) Air pollution from transportation poses a catastrophic threat to climate change and public health. It is imperative to steeply reduce carbon emissions from the transport sector. Discuss the benefits of transport decarbonization.

(2) The transport industry has made great strides with new technologies and policy support. Chart a path for the future transport system in China.

3. Work in groups to make a presentation on any topic related to the theme of the unit.

Vocabulary

Active Reading: A Strategy to Decarbonize the Global Transport Sector

cohort	/ˈkəʊhɔːt/	*n.*	a group of people who share a common feature or aspect of behavior	（有共同特点或举止类同的）一批人
inexorably	/ɪnˈeksərəbli/	*adv.*	unstoppably	不可逆转地；不可阻挡地
anthropogenic	/ˌænθrəpəˈdʒenɪk/	*adj.*	caused by humans or their activities	人类活动引起的
dire	/ˈdaɪə(r)/	*adj.*	very bad	极糟的；极差的
daunting	/ˈdɔːntɪŋ/	*adj.*	seeming difficult to deal with in prospect; intimidating	看上去棘手的；使人畏惧的
couple	/ˈkʌpl/	*v.*	to join together two parts of sth., for example, two vehicles or two pieces of equipment	（把车辆或设备等）连接；结合
adverse	/ˈædvɜːs/	*adj.*	negative and unpleasant; not likely to produce a good result	不利的；有害的；反面的
exclusively	/ɪkˈskluːsɪvli/	*adv.*	only; solely	仅仅；排他地；专门地
substantially	/səbˈstænʃəli/	*adv.*	very much; a lot	非常；大大地

Unit 7　Transport Decarbonization

disproportionate	/ˌdɪsprə'pɔːʃənət/	*adj.*	too large or too small when compared with sth. else	不成比例的；不相称的；太大（或太小）的
stringent	/'strɪndʒənt/	*adj.*	(of a law, rule, regulation, etc.) very strict and that must be obeyed	（法律、规则、规章等）严格的；严厉的
morbidity	/mɔː'bɪdəti/	*n.*	the state of being morbid; the relative incidence of disease	病态；不健全
mortality	/mɔː'tæləti/	*n.*	the number of deaths in a particular situation or period of time	死亡数量；死亡率
deployment	/dɪ'plɔɪmənt/	*n.*	use of sth. (troops, resources, equipment) for a particular purpose so that they are in the right place and ready for quick action	有效运用；部署；调动
upstream	/ˌʌp'striːm/	*adv.*	along a river, in the opposite direction to the way in which the water flows	向（或在）上游；逆流
stakeholder	/'steɪkhəʊldə(r)/	*n.*	a person or company that is involved in a particular organization, project, system, etc., especially because they have invested money in it	（某组织、工程、体系等的）参与方；有权益关系者
electrification	/ɪˌlektrɪfɪ'keɪʃən/	*n.*	the process of changing sth. so that it works by electricity	电气化
synergistic	/ˌsɪnə'dʒɪstɪk/	*adj.*	additionally effective when two or more companies or people combine and work together to achieve their aims	协作的；协同作用的

169

necessitate	/nəˈsesɪteɪt/	v.	to make it necessary for sb. to do sth.	使成为必要
median	/ˈmiːdiən/	adj.	having a value in the middle of a series of values	中间值的；中间的

Further Reading: How Can Airlines Chart a Path to Zero-Carbon Flying?

grapple	/ˈgræpl/	v.	to try hard to find a solution to a problem	努力设法解决
aerodynamics	/ˌeərəʊdaɪˈnæmɪks/	n.	the science that deals with how objects move through air	空气动力学
intrinsic	/ɪnˈtrɪnsɪk/	adj.	being part of the nature or character of sth. or sb.	固有的；内在的；本身的
taxi	/ˈtæksi/	v.	(of a plane) to move slowly along the ground before taking off or after landing	（飞机）（起飞前或降落后在地面上）滑行
eke	/iːk/	v.	to make a small supply of sth. last longer by using or consuming it frugally	（靠节省用量）使……的供应持久；节约使用；竭力维持
dispatcher	/dɪˈspætʃə(r)/	n.	a person whose job is to see that trains, buses, planes, etc. leave on time	（火车、汽车、飞机等的）调度员
discretion	/dɪˈskreʃn/	n.	the freedom or power to decide what should be done in a particular situation	自行决定的自由；自行决定权

Unit 7 Transport Decarbonization

zigzag	/'zɪgzæg/	*v.*	to move forward by making sharp sudden turns first to the left and then to the right	曲折前进
propulsion	/prə'pʌlʃən/	*n.*	the force that drives sth. forward	推动力；推进
soot	/sʊt/	*n.*	black powder that is produced when wood, coal, etc. is burnt	煤烟子；油烟
feedstock	/'fiːdˌstɒk/	*n.*	the main raw material used in the manufacture of a product	原料；给料（指供送入机器或加工厂的原料）
refinery	/rɪ'faɪnəri/	*n.*	a factory where a substance such as oil is refined	炼油厂；制糖厂；精炼厂
fragment	/fræg'ment/	*v.*	to break or make sth. break into small pieces or parts	（使）碎裂；破裂
monoculture	/'mɒnəkʌltʃə(r)/	*n.*	the practice of growing only one type of crop on a certain area of land	单作；单种栽培
gasify	/'gæsɪˌfaɪ/	*v.*	to convert (a solid or liquid, especially coal) into gas	使（固体，或液体，尤指煤）气化
fermentation	/ˌfɜːmen'teɪʃən/	*n.*	the period or event of increasing in size and becoming filled with gas by chemical change, caused by the action of certain living substances such as YEST	发酵
residue	/'rezɪdjuː/	*n.*	a small amount of sth. that remains at the end of a process	剩余物；残留物；残渣

electrolyzer	/ɪˈlektrəʊlaɪzə(r)/	n.	a device that utilizes electricity to split a substance into its basic components	电解剂；电解器
orchestrate	/ˈɔːkɪstreɪt/	v.	to organize a complicated plan or event very carefully or secretly	精心安排；策划；密谋
consortium	/kənˈsɔːtiəm/	n.	a group of people, countries, companies, etc. who are working together on a particular project	（合作进行某项工程的）财团；银团；联营企业
coalition	/ˌkəʊəˈlɪʃən/	n.	a group formed by people from several different groups, especially political ones, agreeing to work together for a particular purpose	（尤指多个政治团体的）联合体；联盟
collaboration	/kəˌlæbəˈreɪʃən/	n.	the act of working with another person or group of people to create or produce sth.	合作；协作
funnel	/ˈfʌnəl/	v.	to cause money, goods, or information to be sent from one place or group to another as it becomes available	传送（资金、商品、信息等）

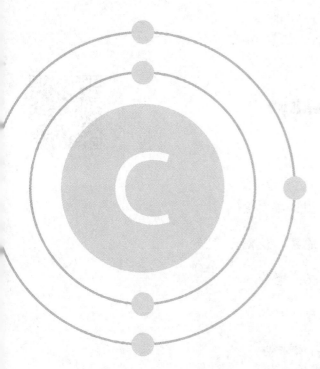

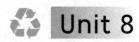 Unit 8

Sustainable Living

We are shaping the world faster than we can change ourselves, and we are applying to the present the habits of the past.

—*Winston Churchill*

Phase I First Sight

Video 1 What Is Sustainability?

New Words

commission /kəˈmɪʃən/ *n.* 考察团；委员会

compromise /ˈkɒmprəmaɪz/ *v.* 危害，使陷于危险

thrive /θraɪv/ *v.* 茁壮成长；兴旺；繁荣

drain /dreɪn/ *v.* （使）排出；滤干

fishery /ˈfɪʃəri/ *n.* 渔业；渔场；水产业

mine /maɪn/ *v.* 采（煤等矿物）

equity /ˈekwəti/ *n.* 公平；公正

1. Watch the video and choose the best answer.

(1) How does the UN define sustainability?

　　A. Development of the society to meet people's own needs.

　　B. Development of the present without sacrificing the future.

　　C. Development of the present to benefit future generations.

　　D. Social development at all cost for the bright future.

(2) Our human life depends on our planet for resources like _____ and_____.

　　A. trees; animals

　　B. soil; minerals

　　C. air; ocean life

　　D. all of the above

Unit 8　Sustainable Living

(3) To develop sustainably, we should _____.

　　A. refrain from consuming the resources from now on

　　B. make the rates of consumption less than those of conservation

　　C. keep steady the levels of replacement and consumption

　　D. find more replenishable resources to meet the needs

(4) We see all the following as problems **EXCEPT** _____.

　　A. disappearing fisheries

　　B. plastics in the ocean

　　C. rising temperatures

　　D. stable levels of resources

(5) According to the framework of the three e's in business, what will be the result if we ignore equity?

　　A. A successful economy is established permanently.

　　B. A few people are affluent while many are hungry.

　　C. A thriving society is developed ultimately.

　　D. The economy prospers and the environment is conserved.

2. Watch the video again and complete the sentences with the words you hear.

(1) Sustainable development is the development that meets the needs of the present without _____ the ability of future generations to meet their own needs.

(2) You can cut down a tree and grow new trees, but that only works if you do it at a rate that these systems can replenish. We call that the _____ rate.

(3) Ultimately sustainability is about understanding that when you make a decision to buy a smartphone, you are impacting someone's life on the other side of the planet who's _____ in mining the materials to put in that smartphone.

(4) One of the other big frameworks that we use to talk about sustainability is the three e's in business, that is, _____, economy, and equity.

(5) If you only look at short-term economic profits, then you're not going to end up with a(n) _____ economy in the long term.

(6) Sustainability is really about _____ all of those and understanding the connections between them.

3. Answer the following questions.

(1) What impacts will individual behavior exert on the environment?

(2) Can you illustrate your understanding of sustainability with examples?

Video II Top Eco-Friendly Cities in the World

New Words

descendant /dɪˈsendənt/ *n.* 后代；后裔

sustain /səˈsteɪn/ *v.* 维持；保持

obligatory /əˈblɪɡətri/ *adj.* 有义务的；强制性的

index /ˈɪndeks/ *n.* 指标；量度

boast /bəʊst/ *v.* 拥有（值得自豪的东西）

gorgeous /ˈɡɔːdʒəs/ *adj.* 美丽动人的

testament /ˈtestəmənt/ *n.* 证明；证据

consistently /kənˈsɪstəntli/ *adv.* 一贯地

cobbled /ˈkɒbld/ *adj.* 铺有鹅卵石的

1. Watch the video and judge whether the following statements are TRUE or FALSE.

(1) _____ The fantastic public transport system helped make San Francisco one of the greenest cities around the globe.

(2) _____ The easily accessible recycling system is the decisive factor that gave Singapore the title of the greenest city in Asia.

Unit 8 Sustainable Living

(3) _____ The Brazilian city of Curitiba achieved the Greenest City Award by developing the biggest bus and public transport systems in the world.

(4) _____ In 2012 Helsinki was named the World Design Capital due to its efficient public transportation systems and its sustainable developments.

(5) _____ Copenhagen will retain No.1 ecologically advanced city in the world by 2030.

2. Watch the video again and fill in the blanks with the words you hear.

Many countries and city governments have been taking action in order to clean up the messes of the past. Ecology has become one of the major (1) _____ in all societies, as we, as a people, have a goal to leave the Earth for our (2) _____ in a better state than when we found it. If we can follow the example set by these Top 10 Eco-Friendly Cities of the world, we'll be well on our way to (3) _____ the Earth for many more generations to come.

San Francisco, United States: This city was the first in the United States to introduce the colored bins recycling system to homes and workplaces, making recycling easy and (4) _____, and more importantly — legally obligatory.

Singapore, Singapore: In addition to the revolutionary public transport system, the number of water (5) _____ in Singapore is remarkably smaller when comparing it with Asia's average index of water waste.

Curitiba, Brazil: Besides high recycling rate, the city's government works hard to make the world a greener place by planting more than (6) _____ trees around the streets and highways.

Helsinki, Finland: In terms of citywide (7) _____, the way the people of Helsinki have maintained their gorgeous city is a testament to their taste and eco-initiative.

Copenhagen, Denmark: Due to its eco-innovation and sustainable employment (8) _____, Copenhagen is consistently ranked as the number one ecologically advanced city in the world.

3. Answer the following questions according to the video.

(1) Which city in the video do you think is the greenest one? Why?

(2) Which cities in China can be considered as eco-friendly cities? Why?

Unit 8　Sustainable Living

Phase II　Getting to Know

Warm-up Activity

Explain the following terms according to what you've explored before class.

- replacement rate
- recycling system
- water wastage
- urban heat island effect

Active Reading

How Cities Are Going Carbon Neutral?

[1] More than half of the world's population currently live in cities, and by the middle of the century, 68 percent of all humans on the planet will live in urban areas. Yet cities are responsible for 60 percent of our greenhouse gas emissions. As urban populations **swell**, so will their impact on the climate.

[2] Cities are also among the places most likely to feel the acute impacts of climate change. The masses of concrete, metal, and glass in urban areas can make them warmer than the surrounding landscape due to the way they absorb, emit, and reflect heat. Water shortages and worsening air pollution threaten to make life in many cities unbearable.

[3] In response, 25 mega-cities have now pledged to become carbon neutral by 2050. These include Rio de Janeiro, New York, Paris, Oslo, Mexico City, Melbourne, London, Milan, Cape Town, Buenos Aires, Caracas, Copenhagen, and Vancouver. If the world hopes to meet its ambition of limiting global temperature increases to 1.5 °C by reaching net-zero carbon emissions by the middle of the century, other cities will almost certainly have to do the same. So, what will our crowded, bustling metropolises have to do to become carbon neutral?

[4] One of the biggest challenges facing cities is their carbon emissions from transport. Some cities are trying to reduce these, along with other types of pollution from vehicles, by introducing Ultra-Low Emission Zones (ULEZ). In London, for example, the area covered by ULEZ has recently been expanded 18-fold from just the city center, and it's now the largest zone of its kind in Europe. The aim is to encourage people to **swap** to lower-emission vehicles, but these still generate carbon emissions during the manufacturing process. To overcome this, some cities are encouraging people to **shun** cars altogether. Paris, for example, is creating 400 miles of new cycle ways and hopes to open up the whole of the city to bicycles by 2026 under a new plan, while the Colombian capital, Bogota, has made 75 miles of streets car-free.

[5] Researchers warn, however, that cultural changes such as getting people to cycle instead of driving can take a long time to realize. But the UN believes that policymakers can use the insight of behavioral science to **nudge** people in the right direction, for example, helping encourage people to cycle more by making it more accessible and easier for them to do so.

[6] Another major contributor to carbon dioxide (CO_2) emissions in cities is the energy required to build, maintain, and run buildings. In 2015, buildings were responsible for 38 percent of global energy-related CO_2 emissions — with the majority produced after construction finishes.

[7] To help reduce the emissions that come from heating, cooling, and powering buildings, however, the construction industry has been making efforts to **incorporate** more alternative energy sources into their design. The aim is to make buildings less reliant on fossil fuels.

[8] In Ulm, southern Germany, the Energon building uses a process called passive heating, drawing on natural energy sources to regulate the building's temperature. Underground canals around the building suck in and heat incoming air in winter, and cool the system in summer with the help of **probes** extending 100 meters (330 feet) underground, where the Earth's natural temperature can be used to cool or heat

Unit 8 Sustainable Living

the air above. This allows the building to use 75 percent less energy for heating and cooling than a standard office building.

⁹ Singapore is often hailed as a leader of sustainable development, but it is hugely dependent on air-conditioning, which is installed in around 99 percent of the city's private homes. Its building sector uses a building rating tool, Green Mark, to encourage sustainability by, for example, using energy-efficient air-conditioning.

¹⁰ But before they're even **inhabited**, buildings have a huge carbon footprint — 11 percent of energy-related carbon emissions are embodied within the construction and materials used. Continuing to construct buildings from concrete and steel could mean emissions reach 600 million tons (544 million tonnes) a year by 2050. At the moment, steel and concrete account for around 16 percent of global CO_2 emissions. But using wood instead could store up to 680 million tons (617 million tonnes) of carbon a year, since trees absorb CO_2 from the atmosphere and using the wood for construction could then lock it away for decades.

¹¹ Cities are also adopting nature-based solutions outside to help store carbon. Medellin in Colombia has planted 30 green corridors along 18 roads and 12 waterways, with 8,300 trees and 350,000 bushes. This has reduced the local temperature by more than 2 °C. Medellin and other dense cities in hot climates can suffer from the urban heat island effect, where hard materials absorb solar energy and pass it back into cities. Nature-based solutions, such as green roofs and living walls, increase biodiversity, have a cooling effect from the **evaporation** of vegetation, and can absorb some particulate pollution in the air.

¹² Green infrastructure has been crucial in Singapore to lower the reliance on air-conditioning. Buildings in the city are being designed to **maximize** natural **ventilation**. Singapore is a very dense city, in a hot and humid climate. And with population growth, the city has a lot of high-rise buildings, facing increasingly warmer temperatures. At the Oasia Hotel in Singapore, for example, a wall of greenery **adorning** the exterior helps cool the building. The temperature of the façade measures 28 °C, compared with the surface temperature of a building **cladded** with metal, which would be around 42 °C.

[13] But cities are only able to make use of their individual resources and climates. Many cities are using this to their advantage — for example, Copenhagen aims to build 360 wind turbines by 2025 to supply most of the electricity demand in the city, while Rio de Janeiro's Museum of Tomorrow uses nearby water from Guanabara Bay to lower the indoor temperature.

[14] But the changes required to reach net-zero carbon require city authorities to balance countless systems and interests — and many operate beyond their control.

[15] Stefan Knupfer who leads sustainable practice for McKinsey says it is important that cities focus on just a few areas needing improvement. In 2017, he helped analyze around 450 tools that were being used to make cities more sustainable, and defined 12 initiatives that cities should focus on to make the biggest difference. These include decarbonizing the electricity grid, optimizing energy efficiency in buildings, offering residents low-carbon transport options and improving how we manage waste. Cities, he says, need to focus on initiatives with the shortest-term impact because time is running out to mitigate the worst effects of global warming. "It's important to work on what we know, not dreaming things up."

(1,115 words)

1. Judge whether the following statements are TRUE or FALSE.

(1) _____ A minority of human population live in urban areas and contribute to more than half of greenhouse gas emissions.

(2) _____ Although it can take a long time to get people to cycle instead of driving, policymakers can turn to science and encourage people in the right direction.

(3) _____ The construction industry has to incorporate alternative energy sources into their design of building, maintaining, and running buildings.

(4) _____ Medellin in Colombia is adopting nature-based solutions outside to help lock carbon away.

(5) _____ Stefan Knupfer advises city authorities to consider countless systems and interests simultaneously to make cities more sustainable in the long run.

Unit 8   Sustainable Living

2. Complete the following sentences with the words or phrases in the box below.

> adorn   swap   mitigate   nudged
> pledged   reliant   metropolis   swell

(1) Modern Shanghai has grown into a highly cosmopolitan _____, spreading far beyond its original location.

(2) We must conserve natural forests so as to _____ the worst effects of global warming.

(3) People can _____ the exterior wall with greenery to help lower the temperature of the building.

(4) There are people who _____ with pride whenever they make some small contributions.

(5) Research showed that humans didn't necessarily have to kill these animals in great numbers, but still _____ them toward extinction.

(6) Many cities have _____ to reach net zero by 2050, the target set by the Paris Agreement on Climate Change.

(7) People may _____ to lower-emission vehicles, but the vehicles still generate carbon emissions during the manufacturing process.

(8) Switching to corn ethanol helps the economy become less _____ on energy sources from other countries.

3. Match the following technical terms with their Chinese equivalents.

I	II
(1) greenery	(　) 供电系统
(2) green mark	(　) 降温效果
(3) ultra-low emission zone	(　) 绿色（环保）标记
(4) powering system	(　) 绿色基础设施
(5) passive heating system	(　) 绿色墙
(6) cooling effect	(　) 绿色植物
(7) natural ventilation	(　) 城市热岛效应

(8) green infrastructure　　　　（　　）超低排放区
(9) living wall　　　　　　　　　（　　）被动采暖系统
(10) urban heat island effect　　 （　　）自然通风

4. Translate the following paragraph into Chinese.

　　Cities are also adopting nature-based solutions outside to help store carbon. Medellin in Colombia has planted 30 green corridors along 18 roads and 12 waterways, with 8,300 trees and 350,000 bushes. This has reduced the local temperature by more than 2 ℃. Medellin and other dense cities in hot climates can suffer from the urban heat island effect, where hard materials absorb solar energy and pass it back into cities. Nature-based solutions, such as green roofs and living walls, increase biodiversity, have a cooling effect from the evaporation of vegetation, and can absorb some particulate pollution in the air.

Phase III Engaging Yourself

Warm-up Activity

Explain the following terms according to what you've explored before class.

* sharing economy   * green living   * green infrastructure

Further Reading

Green Living in China

[1] Green development requires everyone's efforts, and each of us can promote and practice green living. China actively promotes the values and ideas of eco-environmental conservation, raises public awareness to conserve resources and protect the eco-environment, and advocates the practice of a simpler, greener, and low-carbon lifestyle, creating a **conducive** social atmosphere for jointly promoting green development.

Continuing Progress Towards Raising Conservation Awareness

[2] China places particular emphasis on cultivating its citizens' conservation awareness. It organizes systematic **publicity** and other awareness-raising activities in this regard, and advocates a social environment and lifestyle of diligence and frugality. Publicity activities themed on National Energy-Saving Publicity Week, China's Water Week, National Urban Water-Saving Week, National Low-Carbon Day, National Tree-Planting Day, World Environment Day, the International Day for Biological Diversity, and Earth Day, are organized on a regular basis to encourage and persuade the whole of society to engage in green development activities. The idea of eco-friendly living has become widely accepted in families, communities, factories, and rural areas. Material on green development has been

incorporated into China's national education system through **compiling** textbooks on eco-environmental conservation and carrying out education in primary and secondary schools on the condition of national resources including forests, grasslands, rivers and lakes, land, water and grain. Respect for and love of nature have been advocated. Environmental Code of Conduct for Citizens (for Trial Implementation) was published to guide the public to follow a green lifestyle. As a result, a culture of ecological and environmental protection has joined the **mainstream** and been cherished by all.

Widespread Initiatives to Promote Eco-friendly Lifestyles

[3] China has launched initiatives to promote the building of resource-conserving Party and government offices, and develop eco-friendly families, schools, communities, transport services, shopping malls, and buildings, popularizing eco-friendly habits in all areas including food, clothing, housing, transport, and tourism. To date, 70 percent of Party and government offices at and above county level are now committed to resource conservation, almost 100 colleges and universities have realized smart monitoring of water and electricity consumption, and 109 cities have participated in green transport and commutes initiatives. Household waste sorting has been widely promoted in cities at or above **prefecture** level. Much progress is being made as residents gradually adopt the habit of sorting their waste. The Law of the People's Republic of China on Food Waste has been **enacted**, and initiatives have been launched to promote food saving and curb food waste, including a "clean plate" campaign on a large scale, yielding remarkable results as more people are saving food.

Growing Market of Green Products

[4] China has actively promoted energy-saving and low-carbon products such as new-energy vehicles and energy-efficient household appliances. It has provided tax reductions or **exemptions** and government **subsidies** for new-energy vehicles and continued to improve charging infrastructure. As a result, the sales of new-energy vehicles have rapidly risen from 13,000 in 2012 to 3.52 million in 2021. For the seven years since 2015, China has ranked first in the world in the production

Unit 8 Sustainable Living

and sales of new-energy vehicles. In addition, China has steadily improved the certification and promotion system for green products and the green government **procurement** system, implemented an energy efficiency and water efficiency labeling system to encourage the consumption of green products. It has promoted the construction of green infrastructure in the circulation sector such as green shopping malls, and supported new business models such as the sharing economy and second-hand transactions. There is a richer variety of green products and a growing number of people who spend on green products.

Building a Beautiful Home with a Pleasant Living Environment

[5] Urban and rural areas are the carriers of human settlements and activities. China **integrates** the philosophy of green development into urban and rural construction, and promotes beautiful cities and beautiful countryside initiatives. China strives to improve the living environment to build a beautiful home featuring **lush** mountains, green fields, singing birds, and blossoming flowers.

[6] Building beautiful cities featuring harmony between humanity and nature. China has placed great emphasis on urban eco-environmental conservation and has adopted a people-centered approach to urbanization. It has made sound plans for spaces for working, living and eco-environmental conservation, and has worked to make cities livable, **resilient**, and smart. The aim is to build cities into beautiful homes where humanity and nature coexist in harmony. In pursuing urbanization, China respects and accommodates the requirements of nature. It has made use of mountains, waters, and other unique landscapes to integrate cities into nature, so that urban dwellers can enjoy the view and are reminded of their rural origins. Efforts have been made to expand urban eco-environmental space through construction of national garden cities and forest cities, as well as parks and greenways in cities. With increased greenery coverage, the urban eco-environment has been effectively restored. From 2012 to 2021, green coverage of built-up urban areas increased from 39 percent to 42 percent, and the per capita area of park greenery has increased from 11.8 square meters to 14.78 square meters. Great efforts have been made to construct green and low-carbon buildings, and **renovation** of existing ones has been promoted,

contributing to increasingly higher energy efficiency.

[7] Building a beautiful and harmonious countryside that is pleasant to live and work in. Green development is a new driver of rural **revitalization**, and China is exploring new paths for green development in rural areas. It is actively developing new industries and new forms of business such as eco-agriculture, rural e-commerce, leisure agriculture, rural tourism, health, and elderly care, while advancing projects to protect and restore ecosystems. These efforts allow China to approach its goal of having strong agriculture and a beautiful and revitalized countryside. China is continuing to redevelop the whole rural living environment and steadily advancing the construction of modern and livable homes with **sanitary** toilets in rural areas, strengthening the treatment and recycling of domestic waste and **sewage**. As a result, more and more rural areas have access to safe and clean water, paved roads, streetlights, and clean energy. Greater efforts have also been made to protect and utilize traditional villages and carry forward their fine traditions, which have increased their cultural charm. With an improved environment, the vast rural areas have become more sustainable. Lush groves, orchards, and gardens of flowers and vegetables set each other off, and the splendid **pastoral** scene is a treat for the eyes. A beautiful countryside where the skies are blue, the lands are green, and the waters are clear brings people delight with its scenic view.

Conclusion

[8] China has embarked on a new journey to build itself into a modern socialist country in all respects and advance the **rejuvenation** of the Chinese nation. Harmony between humanity and nature is an important feature of China's modernization. The just-concluded 20th CPC National Congress has made strategic plans for China's future development which will help create a better environment with greener mountains, cleaner water, and clearer air. China will keep to the path of green development, continue to build an eco-civilization, and strive to realize development with a higher level of quality, efficiency, equity, sustainability, and security. We will make "green" a defining feature of a beautiful China and allow the people to share

Unit 8　Sustainable Living

the beauty of nature and life in a healthy environment.

(1,213 words)

1. Read the text and choose the best answer.

(1) What has China done to cultivate its citizens' awareness of eco-environmental conservation?

　A. Conservation-awareness-raising publicity activities are organized regularly.

　B. Eco-environmental conservation education has been conducted systemically.

　C. Code of Conduct for Citizens to follow a green lifestyle has been published.

　D. All of the above.

(2) Which of the following is **NOT** eco-friendly?

　A. Resource conservation.

　B. Free water consumption.

　C. Waste sorting.

　D. Green transport.

(3) China has _____ to promote the growth of green product market.

　A. provided tax exemption for new-energy vehicles

　B. supported new business models such as the sharing economy

　C. implemented water-efficiency labeling

　D. all of the above

(4) What is a beautiful home typically featured by?

　A. Harmony between humanity and nature.

　B. Construction of national gardens.

　C. Renovation of existing buildings.

　D. Protection of traditional villages.

2. Replace the underlined parts with the words or phrases in the box below.

> conducive   incorporated   embark on   enact
> be exempted from   rejuvenate   resilience   initiative

(1) The plants in these colonies need tough stems above the soil with <u>special ability to recover from injury</u>, stems that can bent a lot but not break.

(2) The central bank said that creating an environment <u>beneficial</u> to investment was key to reviving national economy.

(3) All households with a disability living allowance claimant will <u>not be affected by</u> this restriction, as would war widows.

(4) The government has yet to <u>start</u> most of the economic deregulation it has long promised.

(5) Alternatively, the government could <u>make into a law of</u> tax credits for people who invest in renewable energy.

(6) The government must not use this <u>new plan</u> as a means of resolving the pension problem through the back door.

(7) With more scholars coming back from overseas, and with the concerted efforts of the whole nation, we have reasons to expect this country will <u>grow lively and robust again</u>.

(8) Auto-makers have <u>included</u> all the latest safety features into the design.

3. Write a summary of the text in 120–150 words. You may use the given words or phrases in the box.

> embark on   awareness   conservation
> energy-efficient products   harmony
> path   launch   strive

Unit 8 Sustainable Living

Phase IV Commitment

1. Translate the following paragraph into English.

"美丽中国，我是行动者"活动在中国大地上如火如荼展开。以公交、地铁为主的城市公共交通日出行量超过2亿人次，骑行、步行等城市慢行系统建设稳步推进，绿色低碳出行理念深入人心。从节水节纸、节电节能、反对餐饮浪费，到环保装修、拒绝过度包装、告别一次性用品，"绿色低碳节俭风"吹进千家万户。简约适度、绿色低碳、文明健康的生活方式正在成为社会新风尚。

2. Work in pairs to discuss the following topics.

(1) List the major features of a green city and discuss them.
(2) Discuss how each business, community, and individual should behave in an environmentally-friendly way.

3. Work in groups to make a presentation on any topic related to the theme of the unit.

Unit 8　Sustainable Living

Vocabulary

Active Reading: How Cities Are Going Carbon Neutral?

swell	/swel/	v.	to increase or make sth. increase in number or size	（使）增加；增大；扩大
swap	/swɒp/	v.	to give sth. to sb. and receive sth. in exchange	交换（东西）
shun	/ʃʌn/	v.	to avoid sb. or sth.	避开；回避；避免
nudge	/nʌdʒ/	v.	to push sb. or sth. gently or gradually in a particular direction	（朝某方向）轻推；渐渐推动
incorporate	/ɪnˈkɔːpəreɪt/	v.	to include sth. so that it forms a part of sth.	将……包括在内；包含；吸收；使并入
probe	/prəub/	n.	a small device put inside sth. and used by scientists to test sth. or record information	探测仪；传感器；取样器
inhabit	/ɪnˈhæbɪt/	v.	to live in a particular place	居住在；栖居于
evaporation	/ɪˌvæpəˈreɪʃən/	n.	the process of becoming a vapor	蒸发
maximize	/ˈmæksɪmaɪz/	v.	to increase sth. as much as possible	使……最大化；增至最大
ventilation	/ˌventɪˈleɪʃən/	n.	a means of providing fresh air	通风

193

adorn	/ə'dɔːn/	v.	to make sth. or sb. look more attractive by decorating it or them with sth.	装饰；装扮
clad	/klæd/	v.	to bond a metal to (another metal), especially to form a protective coating	把两块金属黏在一起（以形成保护层）

Further Reading: Green Living in China

conducive	/kən'djuːsɪv/	adj.	making it easy, possible or likely for sth. to happen	使容易（或有可能）发生的
publicity	/pʌb'lɪsəti/	n.	the work that is done to attract a lot of interest or attention from many people	宣传工作；传播工作
compile	/kəm'paɪl/	v.	to produce a book, list, report, etc. by bringing together different items, articles, songs, etc.	编写（书、列表、报告等）；编纂
mainstream	/'meɪnstriːm/	n.	the ideas and opinions that are thought to be normal because they are shared by most people	主流思想
prefecture	/'priːfektʃə(r)/	n.	an area of local government in some countries, for example, France, Italy, and Japan	（法、意、日等国的）地方行政区域；省；县
enact	/ɪ'nækt/	v.	to pass a law	通过（法律）

Unit 8 Sustainable Living

exemption	/ɪɡˈzempʃən/	n.	official permission not to do sth. or pay sth. that you would normally have to do or pay	免除；豁免
subsidy	/ˈsʌbsədi/	n.	money that is paid by a government or an organization to reduce the costs of services or of producing goods so that their prices can be kept low	补贴（金）；津贴
procurement	/prəˈkjʊəmənt/	n.	the process of obtaining supplies of sth., especially for a government or an organization	（尤指为政府或机构）采购；购买
integrate	/ˈɪntɪɡreɪt/	v.	to combine two or more things so that they work together; to combine with sth. else in this way	（使）合并；成为一体
lush	/lʌʃ/	adj.	(of plants, gardens, etc.) growing thickly and strongly in a way that is attractive; covered in healthy grass and plants	（植物、花园等）茂盛的；茂密的；草木繁茂的
resilient	/rɪˈzɪliənt/	adj.	(of a substance) returning to its original shape after being bent, stretched, or pressed	（物质）有弹性（或弹力）的；能复原的
renovation	/ˌrenəˈveɪʃən/	n.	the state of being restored to its former condition	翻新；整修
revitalization	/ˌriːvaɪtəlaɪˈzeɪʃən/	n.	bringing again into activity and prominence	重振；振兴
sanitary	/ˈsænətri/	adj.	clean; not likely to cause health problems	清洁的；卫生的

sewage	/'suːɪdʒ/	n.	used water and waste substances that are produced by human bodies and that are carried away from houses and factories through special pipes (= sewers)	（下水道的）污水；污物
pastoral	/'pɑːstərəl/	adj.	showing country life or the countryside, especially in a romantic way	田园的；乡村生活的；村野风情的
rejuvenation	/rɪˌdʒuːvə'neɪʃən/	n.	the act or process of making sb. or sth. look or feel younger, more lively or more modern	恢复青春；恢复活力

教师服务

感谢您选用清华大学出版社的教材！为了更好地服务教学，我们为授课教师提供本学科重点教材信息及样书，请您扫码获取。

》 最新书目

扫码获取 2024 **外语类**重点教材信息

》 样书赠送

教师扫码即可获取样书